THE CALL OF THE PRIMES

THE CALL OF THE PRIMES

Surprising Patterns, Peculiar Puzzles, and Other Marvels of Mathematics

Owen O'Shea

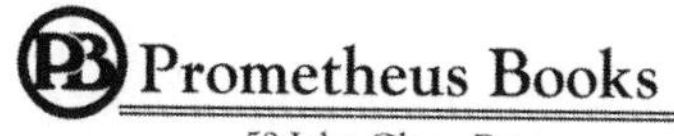
Prometheus Books
59 John Glenn Drive
Amherst, New York 14228

Published 2016 by Prometheus Books

Cover image © Media Bakery
Cover design by Nicole Sommer-Lecht

Inquiries should be addressed to

Prometheus Books
59 John Glenn Drive
Amherst, New York 14228
VOICE: 716–691–0133
FAX: 716–691–0137
WWW.PROMETHEUSBOOKS.COM

20 19 18 17 16 5 4 3 2 1

Library of Congress Cataloging-in-Publication Data Pending

ISBN 978-1-63388-148-8 (paperback)
ISBN 978-1-63388-149-5 (ebook)

Printed in the United States of America

In memory of my late sister, Marie, and her husband, Tim,
and my late brother, Dominic.

CONTENTS

ACKNOWLEDGMENTS

The author, Owen O'Shea, wishes to thank Steven L. Mitchell, editor-in-chief; Jade Zora Scibilia, senior editor; and all of their colleagues at Prometheus Books, for their expert guidance and skill in editing the contents of this book.

INTRODUCTION

Most people, it is probably fair to say, find the subject of mathematics daunting. Perhaps that is because of the way mathematics is taught in our schools. If mathematics is taught in a dull and incomprehensible manner, as I believe it often is, then it is only to be expected that students will find the subject to be equally dull and impenetrable. But if mathematics is taught with emphasis given to its beauty, elegance, and order, then students and the general public will more often than not be tempted to learn more and appreciate it for the fascinating subject that it is. I have long believed that the best way to teach mathematics to anyone is to approach the subject in a spirit of play. Recreational mathematics, in my opinion, is the best tool to do this.

If you can gain a person's interest in a subject and show him that he really can grow actually to like it, then there is a very high probability that he will become good at the relevant subject and even excel in it. The same reasoning applies to the teaching of mathematics. If you can get an individual interested in a good number trick or an interesting number pattern or a mathematical puzzle that perhaps has a very counterintuitive solution, then it is likely that you will foster in that person a strong liking for mathematics; and enjoyment breeds the desire to explore and to seek out new challenges as one's experience increases. Consequently, he or she will most likely develop an equally strong tendency to ask (and hopefully solve) meaningful questions.

It is imperative for mathematicians or scientists not just to attempt to answer questions but also to *ask* questions. The ability to ask meaningful questions is essential for an inquiring mind. Science proceeds and succeeds because human beings ask penetrating questions about how the world operates. Mathematical knowledge expands because mathematicians are curious about the reach and power of their subject to help explain the world and how it works, and in doing so they ask questions about both the steadfast and innovative techniques of mathematics as well as the types of objects (e.g., puzzles, curiosities, and practical applications) that find them learning more and more about this fascinating world we live in. Recreational mathematics helps to foster the ability to ask meaningful questions.

What is recreational mathematics, and what makes it useful? Recreational mathematics is mathematics that is both fun and popular. Its subject matter should be understandable to the interested layperson who may have no training in the field of mathematics. Enjoying recreational mathematics may take the form of solving a good, intriguing mathematical puzzle, often with a surprising and delightful answer, or it may involve learning about a famous mathematical constant such as *pi* or *e*, without the technical language you would expect to find on these topics in a textbook on mathematics. Or it may mean looking at some vaguely familiar subject—like the prime numbers—in a completely new way, and realizing that within their distribution among the infinity of integers there lays an extremely deep mystery that may never be penetrated by human beings.

Recreational mathematics serves two functions. First, there is the entertaining value of the pursuit. Most people, regardless of their upbringing or culture or status in life, like trying their hand at a good puzzle. This is borne out by the fact that for many years, newspapers and magazines around the world have regularly published a mathematical puzzle, often with a prize offered for the first correct solution submitted by a reader.

For twenty-five years, the magazine *Scientific American* ran a monthly column titled "Mathematical Games." It was written by the late Martin Gardner and was hugely popular with the readers of the magazine. Many young readers of the column were so inspired by the articles that they went on to become mathematicians, engineers, and scientists of the first degree. It has been said that when Gardner retired from writing the column in 1981, the sales of the magazine fell steeply as a result. The fact that many engineers, scientists, and mathematicians around the world publicly acknowledge today their debt to Gardner and his column on mathematical games would suggest that recreational mathematics has a huge role to play in the advancement of not only mathematics, but of all scientific knowledge.

The second function of recreational mathematics is the pedagogic one. What better way is there to teach anyone to question the world and all that is in it than to present her with a good mathematical puzzle or trick? In trying to solve the puzzle or in trying to figure out how the trick works, the student will realize that all is not what it seems to be with the puzzle or trick. This in turn will teach her to question things, and to ask why things are the way they are. Could the social order we have be different? Could the laws of nature be different? Could the laws of mathematics be different from what they currently are? Readers will learn that in mathematics—as in life—it is sometimes more rewarding or indeed necessary to look at things from a different angle, to question things, to look for a deeper understanding in order to find the correct solution to a problem. This general approach is essential to solving the many problems the world throws at us as we go through life.

In chapter 2, I make the point that a large part of learning mathematics is about attaining the ability to spot patterns. For example, if you write the first twelve primes in order, you will have the following series: 2, 3, 5, 7, 11, 13, 17, 19, 23, 29, 31, 37. If you inspect this series of numbers, you may spot the fact that—ignoring the first two primes, 2 and 3—every other prime in the series is 1 less or 1 more than a multiple of 6. You may then write the next twelve primes, following 37: 41, 43, 47, 53, 59, 61, 67, 71, 73, 79, 83, 89, 97. If you examine this list of primes, you also find that each prime is just 1 less or 1 more than a multiple of 6.

You may begin to wonder if this is true of every prime, except 2 and 3. Proceeding along this line of thought, the curious investigator soon realizes that every multiple of 6 cannot be prime, because it is divisible by 6. You may then begin to realize that every multiple of 6 is divisible by both 2 and 3. Therefore, any number that is 2 less or 2 more than a multiple of 6 is also divisible by 2, and therefore cannot be prime.

Continuing farther along this train of thought, you might begin to realize that any number that is 3 less or 3 more than a multiple of 6 is divisible by 3, and therefore that number cannot be prime either. Then you would reach the conclusion that the only position a prime can occupy is 1 less or 1 more than a multiple of 6. The first two primes are exceptions to this general rule, simply because they are so close to 1. The only integer less than 2 is 1, so 2 must be a prime, because it has only two different divisors: 1 and 2. (The reader may recall that a prime number [let's call it n] has only two different divisors: 1 and n itself.) Therefore, 2 must be a prime. The only integers less than 3 are 1 and 2, neither of which divide evenly into 3. Therefore, 3 is a prime also.

Because of your ability to spot a pattern in how the primes are distributed, you have unearthed a universal truth about the distribution of prime numbers that is not obvious to the casual observer. Thus mathematicians who can spot patterns lurking behind mathematical data are usually the ones to make new and exciting discoveries.

It is difficult to draw the line between what is generally considered to be recreational mathematics and what many might consider to be serious mathematics. Clearly, what is for some an intriguing puzzle that could be viewed as the mathematical equivalent of unraveling a Rubik's Cube is for others an incomprehensible muddle that merely baffles rather than excites healthy curiosity. Consequently there is often an overlap between the two. For instance, consider graph theory. It is extremely important in the worlds of biology, sociology, engineering, and computer science. Surprisingly, graph theory had its origins in a recreational problem now known universally as *The Seven Bridges of Königsberg*. The problem was eventually tackled and solved by using the basics of what is now known as graph theory by Leonard Euler (1707–1783), one of the greatest mathematicians who ever lived.

Another example of the overlap between recreational mathematics and more serious mathematics is found when you consider the origins of probability theory. The laws of probability theory are used extensively in the modern world. The study of genetics, quality control, and quantum mechanics all depend on this famous branch of mathematics. However, many laypeople do not realize that probability theory had its origins in the sixteenth century when the Italian mathematician Girolamo Cardano (1501–1576) first studied the outcomes of the throws of dice. In 1550 Cardano wrote a book titled *A Book on Games of Chance*, and the infant study of probability was born.

The modern theory of probability was further developed by French mathematician and physicist Blaise Pascal (1623–1662) in his correspondence with Pierre de Fermat (1601 or 1607–1665), the French lawyer and outstanding amateur mathematician. A nobleman and gambler named Chevalier de Méré, who was fond of betting with dice, had found from experience that it was more favorable to bet that one could throw at least one six with four throws of one die rather than throw two sixes once in twenty-four rolls of two dice. This perplexed the gambler, who believed that both bets should have an equal likelihood of happening.

He based his belief on the following reasoning. Consider the one-die game. He knew that the chances of rolling a six in one roll of a single die are 1 in 6 throws (mathematically expressed as 1/6). He then incorrectly reasoned that the chances of rolling a six in the roll of four dice are 4/6, or 2/3.

Now consider the two-dice game. He correctly reasoned that the chance of obtaining a pair of sixes with the roll of two dice is 1/36. Why? Since the probability of obtaining one six with a roll of one die is 1/6, then that one six can be associated with the other six faces that may turn up with the roll of the second die. This means that there are thirty-six different possibilities in which the two dice when rolled may fall. Only one of those thirty-six possibilities will have two sixes showing. Therefore, the probability of rolling two sixes with two dice must be one chance in thirty-six: 1/36.

Chevalier de Méré then incorrectly reasoned that the probability of obtaining a pair of sixes in twenty-four rolls of two dice is twenty-four times greater than the probability of rolling two sixes when rolling two dice. In other words, he believed that the probability of obtaining a pair of sixes in twenty-four rolls of two dice is 24 times 1/6, or 24/36.

This reasoning is clearly incorrect. If it were correct, you could reason that in thirty-six rolls of two dice the probability of rolling two sixes is 36/36, or 1. This would mean it was *certain* to roll two sixes when rolling dice thirty-six times. Of course that conclusion is nonsense.

As a result of his faulty reasoning Chevalier de Méré incorrectly concluded that if one throws a pair of dice twenty-four times, one should have the same chance of obtaining a pair of sixes as if one rolled a single die just four times. He won more often than not when he bet on rolling a six in four rolls of one die. But he found that when he bet on getting a pair of sixes in twenty-four rolls of two dice he lost consistently.

Chevalier de Méré approached Blaise Pascal and asked him to calculate the odds of this dice problem.

Pascal studied the problem. He knew that the probability of *not* rolling a six in one roll of a single die is 5/6. Consequently, the probability of *not* rolling a pair of sixes with four rolls of a single die must be 5/6 multiplied by 5/6 multiplied by 5/6 multiplied by 5/6. This is equivalent to $(5/6)^4$, which in turn equals 625/1296.

Therefore, the probability of *not* rolling a pair of sixes with four rolls of one die is $(5/6)^4$. (In probability theory, an event that is certain to happen has a probability of 1/1, or 1. If there is a one in two chance of an event happening, it has a probability of 1/2. If there are five chances in seven of an event happening, it has a probability of 5/7 that it will happen, and a probability of 2/7 that it will *not* happen.) Therefore the probability of rolling a pair of sixes with four rolls of one die must be $1 - (5/6)^4$. This equals 671 (1296 – 625) chances in 1296, which equals 51.7746 . . . percent. Thus there is more than a 50 percent chance of rolling two sixes with four rolls of one die.

By similar reasoning, one finds that the probability of getting a pair of sixes in twenty-four rolls of two dice is $1 - (35/36)^{24}$, which equals 49.1403 . . . percent. Thus there is less than a 50 percent chance of rolling a pair of sixes with twenty-four rolls of two dice. One probability is just above 50 percent; the other is just below 50 percent. That is why Méré consistently lost on the bet of obtaining a pair of sixes in twenty-four rolls of two dice. It was merely due to the laws of the theory of probability.

It was from such humble beginnings that probability theory grew to the extremely important role it plays today in the modern world.

There are many examples where a recreational mathematical idea, thought to be curious and interesting but of no practical value, turned out to have huge significance in the world of commerce. Consider the binary number system, which is based on the use of 0s and 1s only. It was first discovered in 1679 by the great German mathematician Gottfried Wilhelm Leibniz (1646–1716).

Our decimal number system is based on increasing powers of ten as one goes from right to left. (This is most probably a consequence of the fact that human beings normally have ten fingers (or ten digits). Consider first the number 123. It simply means (from right to left) three times 10^0 (which equals 3); two times 10 (which equals 20), and one times 10^2 (which equals 100). (Any number raised to the power of 0 is equal to 1. Thus 10^0 equals 1.) Therefore the value of 123 in the decimal number system may be easily understood by looking at figure I.1:

100	10	1
1	2	3

Figure I.1

In other words, 123 consists of three times 10^0, two times 10, and one times 10^2. The sum of these numbers is 3, 20, and 100, which equals 123. You will notice that the only digits that appear in the decimal number system are the digits from 0 to 9.

Now in the binary system of numbers, you write numbers in increasing powers of two as you go from right to left. For example, the equivalent of the number 123 in the binary system is 1111011. This is easily understood by looking at figure I.2 below

64	32	16	8	4	2	1
1	1	1	1	0	1	1

Figure I.2

Thus the binary number 1111011 consists of one 2^0 (which equals 1); one 2^1 (which equals 2); zero times 2^2 (which equals 0); one times 2^3 (which equals 8); one times 2^4 (which equals 16); one times 2^5 (which equals 32), and one times 2^6 (which equals 64). Adding these numbers, one obtains $1 + 2 + 0 + 8 + 16 + 32 + 64$. The sum of these numbers is 123. You will probably notice that the only digits that appear in the binary number system are the digits 0 and 1.

For nearly the next three centuries binary numbers were considered to be a mere curiosity. But then electronic computers were invented. It was found that the binary code, with its series of 0s and 1s, was ideally suited to the workings of these new electronic devices. Digital computers store information or perform calculations by electronic circuits being in an "off" or "on" position at any one time. Each of these two possible states corresponds with the two digits, 0 and 1, of the binary code, making the binary numbers a natural choice for the operation of an electronic system.

Although binary numbers did not attract much attention in the Western world, their significance was noted in other parts of the globe. For example, in an article on December 16, 2013, in the prestigious magazine, *Nature*, Philip Ball claimed that the natives in the tiny island of Mangareva in French Polynesia in the Pacific Ocean were using binary arithmetic six hundred years ago. The same article mentions that the *I Ching*, a Chinese text thousands of years old that was known to Leibniz, also contains the binary numbers.

Despite this, it appears that European mathematicians were not greatly interested in binary numbers. That changed, however, in 1854.

In that year, the British mathematician George Boole (1815–1864), discovered a form of algebraic system of logic, which today is known as Boolean algebra. No one, including Boole, thought it would have any practical use. However, in 1937 a young student at the Massachusetts Institute of Technology named Claude Shannon (1916–2001) realized that Boolean algebra was perfectly suited to the workings of an electric circuit. He wrote a master's thesis on his findings in 1937. This thesis was the basis for the use of the binary system in computers. Binary code is used elsewhere also. For example, compact discs, known universally as CDs, make use of the binary code to store data (i.e., words, images, sounds) for future use.

This historical evidence illustrates that a mathematical idea, thought to be merely curious or recreational and not at all useful in the "real world," actually can have huge significance in many practical areas.

Great minds often play with mathematics. In 1939, a group of Princeton University graduate students named Arthur H. Stone, Bryant Tuckerman, John W. Tukey, and Richard Feynman began playing with strips of paper that were folded into the shape of hexagons (think of the shape of a stop sign) and that had their ends glued together. These structures were subsequently named *hexaflexagons* by students at Princeton. The hexaflexagons had the curious property that they could be turned inside out repeatedly, showing different faces that were initially hidden.

Hexaflexagons are very similar to look at from the front as standard hexagons, but within the hexaflexagons are many unexpected structures. As well as offering an increased appreciation of geometry, hexaflexagons may be used to illustrate that sometimes there are tremendous delights lurking behind many mathematical structures that may appear initially to be uninteresting.

At the time, Tuckerman, Stone, and Tukey were graduate students of mathematics at Princeton; Feynman was a graduate student of physics. Tukey was known to teach mathematics to other young graduates, and later he became a professor of mathematics at Princeton. Stone went on to become a lecturer in mathematics at the University of Manchester. Like Tukey, Tuckerman also became a mathematician at the Institute of Advanced Study in Princeton.

Feynman (1918–1988) went on to become one of the best theoretical physicists in the world. It has been said that he looked upon a physics problem in a similar way that he looked upon a mathematical puzzle. Each problem in physics was, to Feynman, a delightful real-world puzzle posed by nature to be solved.

At school Feynman liked recreational mathematics. He was particularly good at finding clever solutions to mathematical puzzles. In his final high school year, he won the New York University Math Championship. I believe it is reasonable to assume that Feynman's interest in recreational mathematics spurred him on in his quest to solve the many physics problems he encountered in later life.

It appears, therefore, that recreational mathematics has a significant role to play in the advancement of human knowledge. The question that now arises is,

Can a mathematical puzzle, trick, or game motivate average interested readers to learn the more serious aspects of mathematics? One of the great German mathematicians certainly believed they could. Leibniz, whom we met earlier and who was one of the two discovers of calculus, said: "I strongly approve the study of games of reason, not for their own sake, but because they help to perfect the art of thinking."[1]

The book that you are now holding in your hands does not claim to contain all entertaining mathematical diversions. Far from it! It is just a small sample from a large collection of topics. Of course, what one reader may consider highly entertaining another may consider only moderately entertaining or perhaps not entertaining at all! However, it is my hope that this collection will have something to please even the most fastidious reader.

I do not claim to be a mathematician. My only claim is that I like writing about mathematics. The fact that I am not a mathematician helps me, I believe, to convey in some small measure the wonder and beauty of mathematics to the interested but nonmathematical reader.

It is my sincere hope that the contents of this book fulfill that aim.

CHAPTER 1

SOME WORDS ON THE *LO SHU* AND OTHER MAGIC SQUARES

Magic squares have fascinated people of all ages through the centuries. A magic square is an array of numbers, usually distinct, arranged in square formation. Thus a magic square contains the same number of horizontal rows as vertical columns. The magic square derives its name from the fact that the sums of the numbers in each row, in each column, and across its two diagonals, are identical. Although no mathematical knowledge is imparted through the study of magic squares, nevertheless many have investigated these squares with the expectation of finding beautiful relationships between the integers within the squares. Those inquisitive investigators—many of whom were amateur mathematicians—have found many beautiful properties! Even great professional mathematicians through the ages have inquired into magic squares, including one of the greatest mathematicians of all time: Leonard Euler.

A *normal* magic square is one in which the integers from 1 to n^2 are arranged in a square pattern so that each row, each column, and each of the two diagonals add up to a specific integer. It is this property that makes the magic square *magical*. The earliest magic square is said to date back to 2800 BCE in China. According to a famous Chinese myth, while walking beside the Lo River, Emperor Yu found a tortoise with a specific pattern on its shell. He named this pattern the *lo shu*. (The word *lo* is the name of the river and the word *shu* means books.) The *lo shu* consisted of a number of dots representing whole numbers that added up to 15 in three horizontal directions, three vertical directions, and two diagonal directions.

Although the belief prevails to this day that magic squares were first discovered well over four thousand years ago, present-day Chinese scholars can only trace its earliest discovery back as far as the fourth century BCE. It is believed today that magic squares migrated from China to India around the fifth century CE and they later spread to Arab cultures. The Arabs used them in practicing astrology and other forms of magic rituals. To this day magic squares are still used in various parts of the world as amulets or good luck charms.[1]

The ancient Chinese attached great significance to the three-by-three magic square. They saw the even numbers as representing *yin*, the female principle, and the odd numbers as representing *yang*, the male principle. The central number 5 represented the earth; 4 and 9 represented metal; 2 and 7 represented fire; 6 and 1, water; and 8 and 3, wood. Thus the four elements were found to exist in the three-by-three square.[2]

The *lo shu* today is usually written as follows:

4	9	2
3	5	7
8	1	6

Figure 1.1

The number of rows or columns of a magic square is said to be its *order*. Thus the three-by-three magic square shown in figure 1.1 is said to be of the order three. Its three rows, three columns, and each of its two diagonals add up to 15, which is said to be the magic square's *constant*. Of course, it is possible to rotate or reflect the *lo shu*. You could, for example, rotate the above magic square such that the right column becomes the top row. The top horizontal row would then be 2, 7, 6, or it could be its reflection (i.e., its reversal), 6, 7, 2; therefore, the middle row could contain the digits 9, 5, 1 or 1, 5, 9, respectively, and the bottom horizontal row could contain the digits 4, 3, 8 or 8, 3, 4. These reflections and rotations, however, do not change the essential nature of the three-by-three magic square. Thus the order-three magic square is unique.

No order-one or order-two magic squares are possible. An order-one magic square can only contain one number and is therefore considered trivial. By convention, mathematicians agree that an order-one magic square does not exist.

Suppose a magic square of order two exists. Substituting letters for numbers, we would have in our square:

AB
CD

Since these four numbers constitute a magic square (where all four digits are distinct) we know that

$$A + B = A + C.$$

Subtracting A from both sides of this equation gives

$$B = C.$$

Therefore, B = C. However all the numbers in a magic square must be distinct. Since our magic square of order two must contain at least two similar numbers, we conclude that an order-two magic square cannot exist.

All order-three magic squares are of the form shown in figure 1.2:

$c - a$	$c + a + b$	$c - b$
$c + a - b$	c	$c - a + b$
$c + b$	$c - a - b$	$c + a$

Figure 1.2

Here, a, b, and c are positive integers such that a is less than b and b is less than $c - a$, and b does not equal $2a$. To obtain the smallest possible order-three magic square, let $a = 1$; $b = 3$, and $c = 5$. Plugging these values in to the magic square shown in figure 2.2, the top horizontal row contains the digits—from left to right—4, 9, 2; the digits in the middle row are 3, 5, 7; and the digits in the bottom row are 8, 1, 6.

The constant of a normal magic square is easily derived. The sum of the integers from 1 to n is $(n\,(n + 1))/\,2$, or $(n^2 + n)/\,2$. The integers that appear in a normal magic square run from 1 to n^2. Thus the sum of the integers in a standard magic square is $n^2\,(n^2 + 1)/2$ or $(n^4 + n^2)/2$. The magic square consists of n rows. Therefore, the sum of each row is $(n^4 + n^2)/2n$, which equals $(n^3 + n)/2$. Thus the constant of a normal magic square is $(n^3 + n)/2$. Hence the constant of a normal three-by-three magic square is $(3^3 + 3)/2$, or 15. The constant of a normal four-by-four magic square is 34; the constant of a normal five-by-five magic square is 65. The series continues with 111, 175, 260, 369, 505, 671, and so on.

Curiously, the *lo shu* magic square just barely exists. In the *lo shu*, there are three rows, three columns, and two diagonals, which each sum to 15. Thus for a normal order-three magic square to exist, eight subsets consisting of triplets of numbers whose sum is 15 must exist in the range of digits from 1 to 9. Fortunately, *exactly* eight such triplets exist. These are: 1,5,9; 2,9,4; 2,5,8; 2,7,6; 3,5,7; 4,3,8; 8,1,6; 6,5,4.

A magic square that includes zero can exist. A unique three-by-three magic square that has a constant of 15 is possible if a zero is included, and if 1 and 9 are excluded. In this case, the three numbers—including zero—that add to 15 must consist of 0, 7, and 8; or 0, 5, and 10. There is no other way that three numbers —if one of them is zero—summed together will equal 15. Therefore, 0 cannot occupy a corner cell, because if it did, it would have to be a part of three triples of numbers that add to 15, and this impossible. Since 5 must occupy the center cell, we find that this magic square must have 10 above the 5 and 0 below it. The numbers 7 and 8 will be on either side of the 0. Thus the top row will consist of 2, 10, and 3; the middle row will contain 6, 5, and 4; and the bottom row will consist of 7, 0, and 8. Of course, an infinite number of three-by-three magic squares that contain multiples of these numbers is possible.

Magic squares of order n, commencing with integer A, may be formed where the numbers within them are increasing in an arithmetic series with a difference of D between terms. The constant of such a magic square may be obtained by using the following simple formula:

$$\frac{1}{2}n\,(2A + (D\,(n^2 - 1))).$$

Thus in the simplest three-by-three magic square the smallest integer is 1, and the integers are increasing in an arithmetic series, with a common difference of 1 between terms. Thus, in the above expression $A = 1$, $D = 1$, and $n = 3$. The expression then produces a constant that equals

$$\frac{1}{2} \cdot 3 \cdot (2 \cdot 1 + (1 \cdot (3^2 - 1))).$$

This equals $\frac{1}{2} \cdot 3 \cdot (2 + 8)$, which equals 15. Therefore, the constant of the simplest three-by-three magic square is 15.

Suppose you begin to construct a three-by-three magic square where the smallest term equals 1, and where the terms are increasing in an arithmetic series, with a common difference of 2 between terms. Thus, in such a square A equals 1, D equals 2, and n equals 3. The above expression then produces a constant that equals

$$\frac{1}{2} \cdot 3 \cdot (2 \cdot 1 + (2 \cdot (3^2 - 1))).$$

This expression equals 27. Thus the constant of a three-by-three magic square, where the smallest integer is 1, and the terms are increasing in an arithmetic sequence with a difference of 2, is 27. Figure 1.3 shows this particular three-by-three magic square.

11	1	15
13	9	5
3	17	7

Figure 1.3

An existing magic square may be used to create a new one. Subtract every number in an order n magic square from $n^2 + 1$, and a new square, called the *complementary* of the first square, is formed.

For example, suppose one forms a three-by three magic square by multiplying each term in the *lo shu* by 30. This creates the following three-by-three magic square.

180	30	240
210	150	90
60	270	120

Figure 1.4

The constant in the three-by-three magic square shown in figure 1.4 is 450, and the value of n is, of course, 3. Therefore, $n^2 - 1$ equals 8. If one subtracts each term in the square in figure 1.4 from 8, one obtains the three-by-three square shown in figure 1.5.

–172	–22	–232
–202	–142	–82
–52	–262	–112

Figure 1.5

Of course, in summing the negative terms in the magic square shown in figure 1.5, you may ignore the negative sign in front of each integer, and you may sum the integers in the magic square as if all the terms are positive integers. You will then find that the constant in this magic square is 426.

If any specific integer is added to or subtracted from each of the numbers in a magic square, a new magic square is formed. The same holds if each of the integers in a magic square is multiplied by a constant specific integer. For instance, consider the four-by-four magic square shown in figure 1.6. The constant in this square is 34. If one adds, say, 12 to each term, the result is the magic square as shown in figure 1.7, which has a constant of 82. On the other hand, if each term in the magic square shown in figure 1.6 is multiplied by some constant, say 23, the result is the magic square that is shown in figure 1.8, which has a constant of 782.

1	2	15	16
13	14	3	4
12	7	10	5
8	11	6	9

Figure 1.6

13	14	27	28
25	26	15	16
24	19	22	17
20	23	18	21

Figure 1.7

23	46	345	368
299	322	69	92
276	161	230	115
184	253	138	207

Figure 1.8

No one yet has discovered a universal formula that determines the numbers of magic squares of a given order, *n*. It is known that there is just one magic square of order three (ignoring reflections and rotations). There are 880 order-four magic squares, ignoring reflections, rotations, and such like. There are 275,305,224 order-five magic squares.[3] Beyond order five the *exact* number of magic squares of order *n* is unknown, although various estimates for *n* up to the value of 12 have been formulated and published.

Many beautiful properties unique to the *lo shu* have been discovered. There are almost certainly many more waiting to be unearthed. I will mention just a few that have been found. The sum of the squares of the integers in the top row is equal to the sum of the squares in the bottom row. The same property holds for the far-left and far-right columns. Consider the digits in each of the three rows as three-digit numbers, reading from left to right. These three-digit numbers are 492, 357, and 816. They sum to 1,665. So also do their reversals: 618, 753, and 294. A similar property applies to the numbers in the columns. It is also curious that $49 - 2 + 35 - 7 = 81 - 6$.

The beautiful order embedded in the *lo shu* also reveals itself when you associate it with the prime numbers, those enchanting integers that are the building blocks of all composites. (A composite number is an integer, greater than 1, that is not a prime number.) Here are a couple of examples of the *lo shu*'s relationship with the primes. Reading the columns in the *lo shu* from the bottom up, the three three-digit numbers 834, 159, and 672 appear. The 834th prime is 6,397; the 159th prime is 937; and the 672nd prime is 5,011. The sum of 6,397, 937, and 5,011 is 12,345, a number consisting of five consecutive digits. Reading the columns from the top down, you encounter the three-digit numbers 438, 951, and 276. The 438th prime is 3,061; the 951st prime is 7,507; and the 276th prime is 1,783. The sum of 3,061, 7,507, and 1,783 is 12,345 plus 6.

Reading across the rows of the *lo shu* from right to left, the following three three-digit numbers appear: 294, 753, and 618. Those three numbers, as well as their reversals, add up to 1,665. The same applies to the three three-digit numbers found in the vertical columns, 438, 951, and 276 (and their reversals). The number of primes less than 1,665 is 261. There are nine cells in the *lo shu*. Curiously, 261 equals 9 times the ninth *odd* prime. Also, 261 = 4 – 92 + 357 + 8 – 16.

Magic squares of various orders often display beautiful patterns when lines are drawn connecting consecutive integers. When such lines are drawn in the *lo shu*, for example, the following pattern emerges:

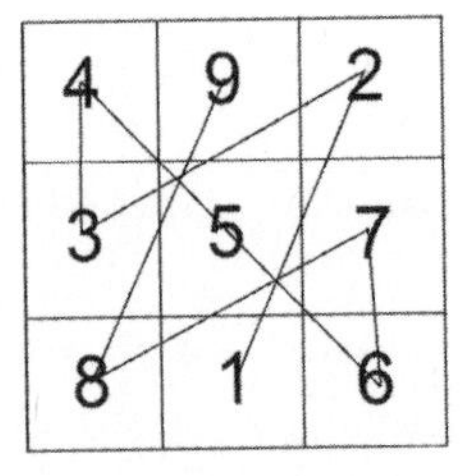

Figure 1.9

In 1984, Dr. Martin LaBar, a professor of science at Southern Wesleyan University, in Central, South Carolina, asked an apparently simple question: Does a magic square of order three exist that contains nine distinct square integers? The late Martin Gardner, the famous writer on recreational mathematics who was mentioned in the introduction, posed the same problem in his column in the magazine *Quantum* in 1996.[4] Gardner remarked in that same column that if a three-by-three magic square of squares exists, which contains nine distinct entries, each of the square numbers within such a square "*are sure to be monstrously large*."

As a consequence of Gardner's article, it is believed that many professional and amateur mathematicians, and a wide assortment of computer programmers around the world, began tackling the puzzle. However, the problem is still unsolved to this day. In 1997, Lee Sallows, a computer programmer in the Netherlands and a widely renowned expert on magic squares, discovered a close miss when he found the three-by-three square shown in figure 1.10, where the nine entries are all perfect squares.[5]

127^2	46^2	58^2
2^2	113^2	94^2
74^2	82^2	97^2

Figure 1.10

Three rows, three columns, and one diagonal of the square shown in this figure all sum to the same constant: 21,609. Unfortunately, the second diagonal sums to 38,307. Sallows published his result in an article in the *Mathematical Intelligencer* in 1997.[6]

Gardner's insight into the difficulty of the problem was confirmed in 1998 when Duncan Buell, of the Department of Computer Science and Engineering at the University of South Carolina, computed that the central cell of a three-by-three magic square of squares would have to be greater than 25 times 10 raised to the twenty-fourth power ($24 \cdot 10^{24}$).[7]

Magic squares of order four are believed to have been first discovered in India around 1000 CE. The constant of a normal four-by-four magic square is $(4^3 + 4)/2$, or 34. Excluding rotations and reflections, the number of four-by-four magic squares is 880. The four-by-four square, shown in figure 1.11, appears in a famous 1514 engraving by Albrecht Dürer (1471–1528), the German engraver and mathematician. The engraving is titled *Melencolia I.*

16	3	2	13
5	10	11	8
9	6	7	12
4	15	14	1

Figure 1.11

The constant of Dürer's square is 34. There are many beautiful properties in the square, but I will mention just a few. The numbers in each of the four quadrants sum to 34: 16, 3, 10, and 5; 2, 13, 8, and 11; 7, 12, 1, and 14; and 9, 6, 15, and 4. The sum of the numbers in the four corners, 16, 13, 4, and 1, sum to 34. The corner numbers in each of the four three-by-three grids within Dürer's square also sum to 34. The four numbers in the central square, 10, 11, 7, and 6, sum to 34. The two central numbers in the top row, 3 and 2, added to the two central numbers in the bottom row, 15 and 14, sum to 34. The same property applies to the two central numbers in the far-left and far-right columns. The two central numbers in the bottom row of Dürer's square give the year in which the engraving was made: 1514. The numbers in the two bottom corners, 1 and 4, give the numeric positions of the initials of Albrecht Dürer.

An order-three magic square consisting of powers of 3, 4, and 5 have been known to be impossible since 1900. Can an order-three magic square consisting of powers greater than 5 exist? The answer is no.[8]

Can an order-four magic square exist in which all of its integers are square numbers? The answer is yes. The first known magic square of any order, consisting of all square numbers, was created by the great Leonard Euler and sent to Joseph-Louis Lagrange in 1770.

Lagrange was born in Turin, Italy, in 1736. He was one of eleven children, of which only two survived into adulthood. Lagrange was mainly a self-taught mathematician, but he was brilliant at the subject. He was appointed professor of mathematics at the Royal Artillery College in Turin when he was only nineteen years old. Lagrange died in Paris, France, in 1813.

The constant of the magic square containing all square numbers that Euler sent to Lagrange is 8,515. This magic square of squares is shown in figure 1.12.

68^2	29^2	41^2	37^2
17^2	31^2	79^2	32^2
59^2	28^2	23^2	61^2
11^2	77^2	8^2	49^2

Figure 1.12

I mentioned earlier that the constant of a normal magic square, where the integers run from 1 to n^2, (the numbers in the square are consecutive) is obtained by the formula $(n^2 + n)/2$, where n is the number of rows in the square. (A magic square that is not normal consists of numbers that do not run from 1 to n^2.) Other formulae or procedures, however, may be used to obtain the constant of any normal magic square. One method uses triangular numbers. The nth triangular number is equal to $n(n + 1)/2$. The series of triangular numbers, commencing with the first, is 1, 3, 6, 10, 15, 21, 28, 36, 45, and so on. T_n is usually used to denote the nth triangular number (e.g., $T_5 = 15$ and $T_8 = 36$, etc.). The following beautiful series gives the constants of successive normal magic squares, beginning with the three-by-three square, then the four-by-four square, then the five-by-five square, and so on:

$$2T_3 + 3 = 15$$
$$3T_4 + 4 = 34$$
$$4T_5 + 5 = 65$$
$$5T_6 + 6 = 111$$
$$6T_7 + 7 = 175$$
$$7T_8 + 8 = 175$$
$$8T_9 + 9 = 369$$
$$\ldots$$

A second method of producing these magic square constants is to write the natural numbers in the form known as *Floyd's triangle*, which is an array of numbers written as follows:

$$1 = 1$$
$$2 + 3 = 5$$
$$4 + 5 + 6 = 15$$
$$7 + 8 + 9 + 10 = 34$$
$$11 + 12 + 13 + 14 + 15 = 65$$
$$16 + 17 + 18 + 19 + 20 + 21 = 111$$
$$\ldots$$

The magic constants of the magic squares of order three, order four, order five, and so on, are produced in the third and consecutive rows of Floyd's triangle.

Finally, I should point out that there are similarities between magic squares (or to be more precise, *Latin squares*—for more on this, see the discussion of figure 1.13 below) and the extremely popular number puzzle known as *Sudoku*.

Numerous newspapers around the world now daily carry a Sudoku puzzle. The word *Sudoku* is the Japanese word for "single number." The premise of a Sudoku puzzle is basically this: You are given a nine-by-nine grid, which consists of nine smaller three-by-three *blocks* or *regions*. At the beginning of the puzzle, various numbers from 1 to 9 appear in various cells within each of the nine three-by-three *regions*. These numbers are usually referred to as the "clues." Most Sudoku puzzles have about twenty-five clues. If they have any more than twenty-five, the difficulty of the puzzle decreases. If they have fewer than twenty-five clues, the complexity of the puzzle increases.

The aim of the Sudoku player is to fill in all of the cells of the nine-by-nine grid with the integers from 1 to 9, so that each of the integers will appear only once in each horizontal row and in each vertical column. In addition, each of the integers from 1 to 9 must appear in every three-by-three region. A good Sudoku puzzle is one in which there is only one unique solution.

There are 5,472,730,538 different Sudoku grids. That's over five billion different arrangements or solutions. This result was first announced in June 2005, and is attributed to Ed Russell and Frazer Jarvis. Jarvis is a senior tutor at the School of Mathematics and Statistics at the University of Sheffield, in England; Russell is an accomplished computer programmer.

The largest number of clues (or starting digits) that can be given in any classic Sudoku puzzle (but which still only allows one unique solution) is seventy-seven.

The smallest number of clues (or starting digits) a Sudoku puzzle can have and still possess a unique solution was long believed by mathematicians (and Sudoku fanatics) to be seventeen. But this had not been proved. Sudoku fanatics around the world reported that when sixteen clues or fewer were given at the start of a Sudoku puzzle, there never appeared to be a unique solution. But when seventeen clues were given, the resulting solution was always found to be unique.

Finally, in 2012, Gary McGuire, Bastian Tugemann, and Gilles Civario in University College Dublin, Ireland, proved that the minimum number of clues was indeed seventeen. They wrote a computer program that tackled the problem in a "brute force" manner, combined with some clever mathematical techniques. The three men used about seven million hours of CPU time running the program before they finally reached the solution.[9]

1	2	3	4
2	1	4	3
3	4	1	2
4	3	2	1

Figure 1.13. A Latin square.

Leonhard Euler studied what are now known as *Latin squares*. The name "Latin square" arose because Euler chose to use Latin characters as symbols in these squares. (Today, numbers or letters of the English alphabet are usually used.) A Latin square is one in which numbers (or letters) are placed in each of the cells inside an n-by-n square, where n is the number of cells to each side of the square. The numbers are placed in the cells so that no number appears more than once in any horizontal row or vertical column. Figure 1.13 illustrates a four-by-four Latin square. Latin squares are sometimes useful in solving specific types of probability problems.

One such problem asks if the four aces, kings, queens, and jacks from a deck of cards can be arranged in a four-by-four array such that no two cards of the same suit or two cards of the same value appear in any row, column, or diagonal. There are 1,152 ways of doing this. Each of these solutions is a Latin square.

A *magic square* is a specific type of Latin square. The difference between the two types of squares is that usually in a magic square you are concerned with the sum of the entries in every row, column, and diagonal. In a Latin square, you are concerned with the ordering of the numbers (or symbols) placed within its cells.

The solution of a Sudoku puzzle results in a Latin square, with the added restriction that the nine squares in a region also contain the digits from 1 to 9.

A Latin square is said to be *normalized* or *reduced* if the symbols or numbers in its top row and extreme left column are in their correct order. There are twelve order-three Latin squares, of which just one is normalized. There are 576 order-four Latin squares, of which there are only four that are normalized. There are 161,280 Latin squares of order-five, of which 56 are normalized. As the numerical order of a Latin square increases, the total number of Latin squares increases at a furious rate. There is no known formula for calculating the exact number of Latin squares of order n, when n increases in size beyond 11. These values of n are available as sequence A002860 on the On-Line Encyclopedia of Integer Sequences; the OEIS Foundation was founded in 1964 by N. J. A. Sloane.[10]

While writing this chapter, I asked my friend, Dr. Cong, the well-known Chinese numerologist and number expert who lives (at present) on the West Coast of the United States, if he had any comments to make concerning the *lo shu*.[11] He, naturally, replied that he had. Dr. Cong subsequently conveyed the following information to me in an e-mail:

Consider the top row of the *lo shu*. The digits in the top row are 4, 9, and 2. The constant of the three-by-three square is 15. The fifteenth prime is 47, which equals 49 minus 2. From 49 objects, the number of ways of selecting two objects, ignoring their order, is 1,176, which equals 4 times 294. That last number is 492 reversed.

There are 94 primes that are less than 492 and 141 primes that are less than 816. The difference between 141 and 94 is 47, which is the fifteenth prime. (Fifteen is, of course, the constant of the *lo shu*.) The sum of 94 and 141 is 235, which is 5 times the fifteenth prime. The number of primes less than 235 is 51, a number that contains the digits of 15 in reverse. The product of 94 and 141 is 13,254. That number contains five consecutive digits.

The following equations that Dr. Cong gave me are surprising:

$$492^2 + 357^2 + 816^2 = 618^2 + 753^2 + 294^2$$
$$438^2 + 951^2 + 276^2 = 672^2 + 159^2 + 834^2$$

The fact that 492 + 357 + 816 (or their reversals) equals 1,665 and that 438 + 951 + 276 (or their reversals) also equals 1,665 may appear to be insignificant. However, there are 261 primes that are less than 1,665. The number 5 occupies the central cell in the *lo shu*. The sum of the first five squares is 55. There are 55 primes that are less than 261.

Dr. Cong also pointed out that the number 492 (which appears in the top row) equals 123 times 4, and that half of the number 816 (which appears in the bottom row) equals 12 times 34. He also gave the following curiosity involving the number 492 (it appears in the top row of the *lo shu*); the numbers 3 and 57 (they appear in the middle row); and the number 816 (it appears in the bottom row):

$$1368 = 492 + 3 + 57 + 816 = 123 + 456 + 789.$$

The Chinese number buff also mentioned that 666, the famous numeric sign of the beast (in the Bible), is lurking within the *lo shu*. To find it, sum the following:

$$4^3 + 9^2 + 2^1 + 8^3 + 1^2 + 6^1 = 666.$$

Dr. Cong also included the following equation:

$$4^3 + 3^2 + 8^1 + 2^3 + 7^2 + 6^1 = 144.$$

The Chinese doctor suggested that perhaps the numbers in the *lo shu* were hinting at the existence of the first 144 decimals of pi, long before they were discovered. Why? Because, as Dr. Cong pointed out, it was curious that the first 144 decimal digits of pi summed to 666! Pi is the ratio of a circle's circumference to its diameter. If the diameter of a circle is 1 unit, the length of its circumference is equal to pi units. The number known as pi equals 3.1415 926535897932384626433832 79 . . . . The decimal expansion goes on forever. It has been calculated to trillions of decimal places in recent years by computer programmers. Mathematicians have not found any *apparent* pattern in its decimal expansion. Pi is one of the most famous numbers in the whole of mathematics; it is discussed later in this book.

Dr. Cong mentioned that one of the great experts on magic squares and cubes was the late John Hendricks. Hendricks was born in Canada on September 4, 1929, and he died on July 7, 2007.

In 1951, Hendricks obtained a degree in mathematics from the University of British Columbia, in Canada. He was employed by the Canadian Meteorological Service for thirty-three years and took early retirement in 1984.

John Hendricks became interested in magic squares and cubes when he was just thirteen years old, and he began collecting different specimens. He had a natural flair for the subject and subsequently made major discoveries in this field. Later in life, he wrote extensively about his discoveries and the methods he used in unearthing such astounding number curiosities.

Dr. Cong said that there were a number of curious correlations between the numbers in the *lo shu* and significant numbers that cropped up in John Hendricks's life. For example, the good doctor mentioned that Hendricks was born on the fourth day of the ninth month. Those two integers (4 and 9) appear from left to right on the top row of the *lo shu*. He was born in the twenty-ninth year of the century. That number (29) appears from right to left in the top row. He died on July 7, which was the 188th day of the year. The number 188 equals 2 times 94, numbers which are found on the top row, reading from right to left. Hendricks was seventy-seven years and 306 days old when he died. The numbers 4, 5, and 6 appear in one diagonal in the *lo shu*. The sum of the squares of these numbers is 77. The number 306 equals 49 + 235 + 7 + 8 + 1 + 6. These numbers appear in the *lo shu*, reading from left to right and from top to bottom.

Dr. Cong mentioned that John Hendricks was born on the 3,169th day of the decade. The number 3,169 equals $4 + 92 \cdot 35 - 7 - 8 \cdot 1 \cdot 6$. Those are the digits that appear in order from left to right and from top to bottom in the *lo shu*.

Dr. Cong pointed out that John Hendricks died on the 2,379th day of the century. Curiously, 2,379 equals $4 \cdot 9 + 2357 - 8 - 1 \cdot 6$.

Finally, the good doctor mentioned that the number 7 appeared to be associated with the life of Hendricks. For instance, he was born on the fourth day of the ninth month. If the digits 4 and 9 are placed side by side, they form the number 49, which equals 7^2. John Hendricks died on the seventh day of the week on July 7 (7/7) in the seventh year of the (7 + 7 + 7)th (twenty-first) century. He was seventy-seven years old when he died.

CHAPTER 2

THE CALL OF THE PRIMES

The prime numbers are those strange, enchanting integers that have only two divisors, themselves and 1. The sequence of primes begins 2, 3, 5, 7, 11, 13, 17, 19, and so on. As we go up the numbers, the primes seem to become fewer and fewer. Do we ever reach the point where the largest prime is found? The answer is no. Two thousand years ago, the Greek mathematician Euclid proved that the number of primes is infinite. No matter how far we go along the number line, we will always find new prime numbers.

Many interesting questions concerning primes have been asked over the years. Two of these questions are: Is there always at least one prime between the squares of two successive integers? And are there are an infinite number of pairs of twin primes?

Consider briefly the first question. The prime number 3 lies between 1^2 and 2^2. The primes 5 and 7 lie between 2^2 and 3^2. The primes 11 and 13 lie between 3^2 and 4^2. But whether there is always at least one prime between successive square integers is still unknown.[1]

The second question relates to twin primes. Twin primes are 3 and 5, 5 and 7, 11 and 13, 17 and 19, 29 and 31, and so on. They are named "twin" primes because only the number two separates the pair of primes in question. Are there are an infinite number of pairs of twin primes? Or is there some large number, N, where beyond that number there are no more twin primes? This is called the *twin prime conjecture*. Although some of the top mathematicians over the years have worked on this problem, no proof that pairs of twin primes are infinite has been found.

Most mathematicians, it appears, believe that the conjecture is true. However, a very small group of professional mathematicians have contemplated the possibility that the twin prime conjecture may be false. That would explain, they argue, our inability to find a proof of the conjecture!

A major advance in the attempts to solve the twin primes conjecture was made in 2013. A lecturer named Yitang Zhang at the University of New Hampshire in the United States submitted a paper to a journal called the *Annals of Mathematics*. The paper claimed to have made a startling discovery in relation to the twin primes conjecture. Zhang was apparently able to prove that there is some integer, N, which is less than seventy million, such that there are an infinite number of pairs of primes that differ by N.[2] The paper was checked by a number of competent referees who appeared satisfied that the mathematics in it is rock solid.

This was a tremendous leap forward in the quest to find a solution to this famous problem. What Zhang was able to prove was that no matter how far out we go on the number line—even to primes containing billions of digits—we will always be able to find pairs of primes that differ (call this difference $2k$) by less than seventy million. Seventy million may seem like a very large number, but it is a lot less than infinity!

Many mathematicians immediately hoped that this proof would mark a starting point at which subsequent proofs might reduce the figure of 70 million all the way down to 2, thus finally proving the twin prime conjecture.

Within a month of Zhang's proof being published, other mathematicians had succeeded in reducing the value of $2k$ to less than 60 million; in a series of other proofs, it was reduced further still, to just over five thousand.[3] Since then, a postdoctoral researcher named James Maynard at the University of Montreal seems to have pushed this gap of $2k$ down to six hundred.[4]

Can the gap be reduced to 2? That would be the universal hope in the mathematical community. However, mathematicians think that the methods used by Zhang and Maynard may not be sufficient to do the trick—that some other, perhaps new, mathematical technique may be required to achieve that. The mathematical community around the world, it seems, holds its breath to see if the problem can be finally conquered.

Probably the most famous of the problems concerning primes is known as *Goldbach's conjecture*. Christian Goldbach, a Russian mathematician, asked the great Swiss mathematician Leonard Euler in a letter in 1742 if every even number greater than two can be expressed as the sum of two prime numbers. Euler, one of the greatest mathematicians who ever lived, couldn't prove Goldbach's conjecture. Neither could any other mathematician since that time. Every even number that has been checked to date can be expressed as the sum of two primes. But is it true for *all* even numbers? No one knows!

Most mathematicians believe the conjecture is true. But in spite of the fact that much work has been done on the conjecture, a proof, if it exists, has eluded the greatest mathematicians since Euler's day.

All even numbers up to $2 \cdot 10^{17}$ have been checked by computer, and every one of those even numbers can be expressed as the sum of two primes. However, this does not strengthen the conjecture in any significant way, because the number of integers up to $2 \cdot 10^{17}$ is only a very tiny proportion of the infinite number of integers. For all we know, there may be an even number somewhere beyond 10^{100}, for instance, that cannot be expressed as the sum of two primes. Maybe it is the only such even number. Maybe there are a limited number of them. Or perhaps there are an infinite number of them scattered far out along the number line. We just don't know.

Much progress on Goldbach's conjecture has been made over the years. In 1937, the Russian mathematician Ivan Matveyevich Vinogradov (1891–1983) proved that every sufficiently large odd number can be expressed as the sum of three primes. This result is now known as *Vinogradov's theorem.*[5]

I should point out here that a mathematical theorem is a result that has been *proved.* On the other hand, a mathematical conjecture is a guess. That guess, or conjecture, may or may not be proved at some point in the future. But until it is proven, the guess is known as a "conjecture."

I should also point out that in a mathematical proof, mathematicians can often prove something extremely interesting about integers beyond a certain integer (let's call that certain integer N) without determining the value of N. They will then describe N in the particular proof as being a "sufficiently large number." This happens frequently in mathematical proofs concerning number theory.

In 2013, the result obtained in Vinogradov's theorem was improved upon by the Peruvian mathematician Harald Helfgott. Helfgott was able to prove that every odd number greater than 5 is the sum of three primes. This was a major breakthrough in getting a step closer to the truth or otherwise of Goldbach's conjecture.[6]

Back in 1973, another breakthrough in Goldbach's conjecture was made. The Chinese mathematician Chen Jingrun (1933–1996) proved that every even number is the sum of a prime and a semiprime. (A semiprime is a number that is the product of two primes.) For example, $40 = 2 + (2 \cdot 19)$; $76 = 11 + (5 \cdot 13)$; $504 = 2 + (2 \cdot 251)$; and $1{,}080 = 7 + (29 \cdot 37)$. Chen's result was another major milestone along the road toward a solution to Goldbach's conjecture.[7]

In 2002, another giant step forward was made concerning Goldbach's conjecture. The Chinese mathematician Ying Chun Cai, of Tongji University, Shanghai, proved that there is a natural number (call it *N*) such that every even number greater than *N* (call this larger number *Y*) is equal to a prime number less than *Y*, plus a semiprime. The proof does not tell us the value of *N*. It only tells us that such a number as *N* exists.

Nonmathematical readers may be asking at this stage what the point of all these conjectures is in relation to the primes. The answer lies in the desire of mathematicians to search for knowledge. To a mathematician, the prime numbers lie at the heart of mathematics. Yet there does not appear to be any great degree of order to their distribution.

That is extremely disappointing to mathematicians. Mathematics is about order and logic. Yet the distribution of the primes along the number line appears to be down to mere chance. For example, if you are informed that the thirtieth prime is 113, there does not appear to be any method (except to actually search) for you to find the next prime. This is unsatisfactory from the mathematician's point of view.

Mathematicians know that mathematics is a powerful tool to describe the order of the natural world. Consequently one would not expect that at the very heart of mathematics there appears to be little or no order.

Mathematicians feel it in their bones that nature would not have left the distribution of the primes to mere chance. Saying the same thing another way, there is a deep philosophical belief among mathematicians that there must be some simple, beautiful order to the way the primes are distributed.

The prime numbers are the building blocks of arithmetic. Every number is either a prime number or a number that equals the product of primes. For example, 29 is a prime number. Its only divisors are 1 and 29 itself. On the other hand, 42 is not a prime number. It is divisible by 1, 2, 3, 6, 7, 14, 21, and 42. Therefore, 42 is a composite number. Note, however, that 42 is the product of 2, 3, and 7. Those are three prime numbers. So, in a sense, prime numbers are the atoms that all composite (nonprime) numbers are made of.

The pure mathematician studies mathematics because she finds the subject interesting and beautiful. She is intrigued by the fact that mathematical proofs are universal and timeless. In other words, if something in mathematics is proved in the United States one hundred years ago, then that proof will also hold today in China, or anywhere else on Earth. In fact, that proof will also be true a thousand years from now, or a million years from now.

The proof will always be true! This is one of the great attractions of mathematics. Ordinary mortals like you and I, whose ancestors lived in the trees just a few million years ago, can discover deep mathematical truths that are timeless and eternal. Isn't that strange! Isn't that wonderful!

The pure mathematician is not interested or concerned in whether there are practical applications to the discoveries made in mathematics. He studies mathematics because of the beauty he finds inherent in the subject. He feels as though he is an explorer. But the world he explores is not the physical universe, but rather the abstract world of mathematics. Nevertheless, that world appears to the mathematician to be just as real as the physical world.

Indeed, the world of mathematics appears to be more real than the physical universe. Long after the physical world has disappeared trillions of years from now, many mathematicians believe that all mathematical theorems will somehow still exist and will still be true. Even when the universe is no more, the fact that 1 plus 1 equals 2 will still be true.

The applied mathematician, on the other hand, is interested in applying the results of mathematical proofs to various projects in the physical world. For example, applied mathematicians make very practical use of the prime numbers by using them in various codes. Every time someone purchases some item online, for instance, the primes are used to maintain the security of credit cards of the purchaser.[8]

We haven't the space to go into detail here on how exactly this highly complicated coding system is achieved or operated. But we will say this: The code used to safeguard Internet commerce is known as the RSA algorithm, in honor of the two computer scientists Ronald L. Rivest and Adi Shamir, and the mathematician Leonard Adleman, who invented the algorithm in the late 1970s.

Essentially, the security of commerce transacted on the Internet is based on this fact: There is no known method of relatively quickly factoring a large number that equals the product of two prime numbers, consisting of say, two hundred digits.[9]

I mentioned earlier that numbers that are equal to the product of two primes are known as *semiprimes*. Suppose I send you a large semiprime that consists of six hundred or seven hundred digits. I ask you to factor that number. You will find it next to impossible to find the two prime factors. Why? Because mathematicians do not know of any method of factoring such large numbers in a relatively short period of time. Even if supercomputers are used to factor that large semiprime, it would still take thousands of years to find the two prime factors!

The mathematical community is satisfied that there is no known method to factor large semiprimes relatively quickly. However, if some mathematician (or group of mathematicians) figured out a way to factor six-hundred-digit semiprimes relatively quickly, the consequences would put an end to all commerce on the Internet.[10]

The world depends on mathematics to a huge degree. For example, computers would not exist if it were not for mathematics. CD players and CDs would not be possible without mathematics. Neither would air travel. Air travel is possible only because we know the mathematics of air flow and of control systems. Statistics is essential in the study of medicine, for analyzing data in relation to diseases or the effectiveness of drugs. And so on.[11]

I mention all of this to inform you that prime numbers are indispensable in the modern world. But beside this, the search for patterns within the distribution of prime numbers raises a major question about the nature of mathematics.

That question has been pondered by mathematicians and philosophers for decades. The question—which is still unanswered—is, Why is mathematics so accurate in describing the natural world? Albert Einstein, who believed that mathematics was invented rather than discovered, phrased the question as follows: "How can it be that mathematics, being after all a product of human thought which is independent of experience, is so admirably adapted to the objects of reality?"[12]

It is one of the world's great mysteries that discoveries made by pure mathematicians that initially appear to have no practical use in the physical world, often turn out (sometimes years later) to be of tremendous use and benefit to the workings of technology. No one knows why this is the case. If there is a lesson to be learned from all of this, it is that we should be hesitant to state what is or is not useful or practical in mathematics.

For example, who would have thought that there is a link between prime numbers and the number pi? Pi equals 3.14159265 . . . . Its decimal expansion goes on forever. (We will discuss pi in more detail later in this book.)

Mathematicians use the symbol π (pronounced "pie") to represent the number 3.14159265 . . . . If the diameter of a circle is one unit, then the length of its circumference is 3.14159265 . . . , which equals π.

Here's an equation illustrating how π and the prime numbers are linked[13]:

$$\frac{\pi^2}{6}=\frac{2^2}{2^2-1}\cdot\frac{3^2}{3^2-1}\cdot\frac{5^2}{5^2-1}\cdot\frac{7^2}{7^2-1}\cdot\frac{11^2}{11^2-1}\cdot\frac{13^2}{13^2-1}\cdot\cdots$$

When you reflect upon this equation, you might note that it is remarkable that the prime numbers should be so intimately related to the circumference of a circle that has a diameter of one unit. In contemplating the question, it is a normal human response to ask what have prime numbers to do with circles.

All of this investigation into the primes may appear to be for the interested mathematician only. It may seem that the primes have little to do with the real world. However, a startling claim about primes has been made in recent years.

Strong arguments have being made by some physicists and mathematicians that the distribution of the prime numbers appears to be very intimately related to the energy levels in the nucleus of a heavy atom when it is repeatedly struck by low-level energy neutrons. (If this is true, it will be further evidence that mathematics is intimately related to nature.[14])

Over the centuries, mathematicians noticed that there is a prime between 2 and 4; that there is at least one prime between 3 and 6, between 4 and 8, and between any integer they inspected and double that integer. They wondered if this is generally true: Is there always at least one prime between an integer, N, which is greater than 1, and twice that integer, $2N$?

The answer is yes, there always is. The question was first proposed by the French mathematician Joseph Bertrand (1822–1900). It became known as *Bertrand's postulate*. It was eventually proved in 1850 by the Russian mathematician Pafnuty Chebyshev (1821–1894). Consequently, Bertrand's postulate became known as *Bertrand-Chebyshev theorem.*[15]

In 1998, the American mathematician Nathan Jacob Fine (1916–1994) used the following little jingle to sum up the result of Bertrand-Chebyshev theorem: "Chebyshev said it, and I'll say it again; there's always a prime between *n* and 2*n*."[16]

The fact that there is always at least one prime between any integer greater than one and double that integer tells us that the primes go on forever. The fact that the primes are infinite is mysterious in itself. We know that the integers (1, 2, 3, 4, 5, . . .) are infinite. But there appears to be fewer primes than there are integers. For example, within the first 100 integers, 25 are prime. Within the first 1,000 integers, only 168 are prime. Within the first 10,000 integers, only 1,229 are prime. Yet it can be proved that the number of integers and the number of primes are the same. They are both infinite!

Mersenne primes, named after the French monk, philosopher, and mathematician Marin Mersenne (1588–1648), are primes of the form $2^p - 1$, where p is a prime. The forty-seventh Mersenne prime was discovered on August 23, 2008, by Edson Smith. That prime equals $2^{43112609} - 1$. It contains 12,978,189 digits.[17]

The largest known prime number at the time of writing (October 2015) is $2^{57885161} - 1$. It is also a Mersenne prime. It is, in fact, the forty-eighth Mersenne prime discovered thus far. It contains 17,425,170 digits, and was discovered on January 25, 2013, by Dr. Curtis Cooper, an American mathematician at the University of Central Missouri, as part of the Great Internet Mersenne Prime Search (GIMPS).[18]

Does the series of Mersenne primes continue forever? No one knows. Most mathematicians believe that it probably does, but a proof of this has not yet been found.

I mentioned in the introduction how important it is for a mathematician to spot patterns. If a mathematician studies the distribution of the primes, for instance, he would notice that the product of the first two primes, 2 and 3, is 6; 6 plus 1 is 7. He might then notice that the next prime is 11 which is 5 more than 6, and 5 is a prime. The mathematician might then calculate the product of the first three primes, 2, 3, and 5, obtaining 30, and add 1 to that, obtaining 31. The next prime is 37, which he may notice is 7 more than 30, and 7 is a prime number. The mathematician may then feel he is on to something. He will wonder if this is always true. It is this ability to spot patterns that others have missed that sometimes can lead to new results in mathematics.

Reo Franklin Fortune (1903–1979) was probably thinking along these lines when he made what appears to be a new discovery about the distribution of the primes. Fortune was a lecturer in anthropology in Cambridge University, in England. He is known to the mathematical community because of a conjecture he made concerning prime numbers, which is now known as *Fortune's conjecture*.[19]

The conjecture involves prime numbers of the form $p_n\# \pm 1$, where $p_n\#$ is the primorial of p_n (that is, the product of the first n primes). For example, $p_2\# \pm 1 = (2 \cdot 3) + 1 = 7$, or $(2 \cdot 3) - 1 = 5$; $p_3\# \pm 1 = (2 \cdot 3 \cdot 5) + 1 = 31$, or $(2 \cdot 3 \cdot 5) - 1 = 29$; $p_4\# \pm 1 = (2 \cdot 3 \cdot 5 \cdot 7) + 1 = 211$, or $(2 \cdot 3 \cdot 5 \cdot 7) - 1 = 209$. And so on.

Numbers of the form $p_n\# \pm 1$ sometimes produce primes, but not always. Let p equal the product of the first n primes. Now consider the first prime greater than $p + 1$. Let that prime equal q. Subtract p from q. Fortune conjectured that the result, which is now known as a *Fortune number* (some mathematicians describe it as a *Fortunate* number), will always be prime.

For example, start with the first prime, which is 2. Now consider the first prime greater than (2 + 1). The first prime greater than (2 + 1) is 5. Subtract 2 from 5. The answer is 3, which is a prime. Therefore, 3 is the first Fortune prime. Consider the product of the first two primes: $2 \cdot 3 = 6$. The first prime greater than (6 + 1) is 11. Subtract 6 from 11. The answer is 5, which is a prime. Therefore, 5 is the second Fortune prime. Consider the product of the first three primes: $2 \cdot 3 \cdot 5 = 30$. The first prime greater than (30 + 1) is 37. Subtract 30 from 37. The answer is 7, which is a prime. Therefore, 7 is the third Fortune prime. Consider the product of the first four primes: $2 \cdot 3 \cdot 5 \cdot 7 = 210$. The first prime greater than (210 +1) is 223. Subtract 210 from 223. The answer is 13, which is a prime. Therefore, 13 is the fourth Fortune prime. And so on. Fortune conjectured (guessed) that this procedure would always produce a prime. Hence the name *Fortune's conjecture.*

The first fifteen Fortune numbers produced by the above procedure are 3, 5, 7, 13, 23, 17, 19, 23, 37, 61, 67, 61, 71, 47, and 107. These numbers are all primes.

Is Fortune's conjecture about the primes true? No one knows. It is, after all, a guess; therefore, it is known as a *conjecture*. If it is eventually proved, it will be known as a *theorem*. However, it can be shown by mathematical reasoning that it is *likely* that Fortune's conjecture is true. At the time of writing this chapter (October 2015), the first two thousand Fortune numbers have all been verified to be prime.[20]

It was the great German mathematician Carl Friedrich Gauss (1777–1855) who first detected a hint of order in the manner in which the primes are distributed. At the tender age of fifteen, Gauss discovered a theorem that gives approximate estimates on how primes are distributed among the integers. The theorem is now known as the *prime number theorem* (PNT). The PNT was later proved independently, in 1896, by the French mathematician Jacques Hadamard (1865–1963) and the Belgian mathematician Charles-Jean de la Vallée Poussin (1866–1962).[21]

The distribution of prime numbers has fascinated mathematicians for over two thousand years. Among relatively small numbers primes seem to abound. For example, 40 percent of numbers below 10 are prime. But as we go up through the numbers, the percentage of them that are prime decreases. Twenty-five percent of numbers under 100 are prime, and 16.8 percent of numbers under

1,000 are prime, while 0.078498 percent of numbers less than 1,000,000 are prime. (The 78,498th prime is 999,983.) Hence, on average, the gap between successive primes seems to get larger and larger as we go up the numbers.[22]

But there are exceptions. Every so often we come across successive primes that differ by just 2. For instance, we find that 101 and 103 are twin primes. So are 1,019 and 1,021; and 10,037 and 10,039. This appears to indicate that the distribution of primes is not completely predictable and may be down to mere chance. But mathematicians can prove—using the prime number theorem—that there is *some* order to the way the primes are distributed. But what precisely that degree of order is is unknown.[23]

The mathematician George F. Simmons quoted Leonard Euler's attitude toward the primes: "Mathematicians have tried in vain to this day to discover some order in the sequence of prime numbers, and we have reason to believe that it is a mystery into which the mind will never penetrate."[24]

Was the great Euler correct when he said that the deep mystery of the distribution of the primes will never be understood by human minds? Or will the day come when this deep, tantalizing, and ancient mystery will be cracked by some modern-day Euler?

I asked my good friend Dr. Cong if he had any comments to make on the primes. He sent an e-mail containing the following notes.

> Hi, Owen,
>
> One curious fact concerning the distribution of the primes is that we can find gaps as large as we like between consecutive primes. For example, consider the number 100 factorial, which is usually written as 100! [The factorial of an integer is the product of a positive integer and all those below it. For example, the factorial of 3 is $3 \cdot 2 \cdot 1$. Therefore, the factorial of 3 equals 6. Instead of saying "the factorial of 3," mathematicians usually state "3 factorial." This is usually written as "3!"] The number 100 factorial equals $1 \cdot 2 \cdot 3 \cdot 4 \cdot 5 \cdot \ldots \cdot 97 \cdot 98 \cdot 99 \cdot 100$.
>
> Now 100! is obviously divisible by 2, since 2 is one of its factors. Therefore, 100! + 2 is also divisible by 2. And 100! is also divisible by 3, since 3 is one of its factors. Therefore, 100! + 3 is also divisible by 3. For similar reasons, 100! + 4 is divisible by 4; 100! + 5 is divisible by 5; and so on, up to and including 100! + 100, which is divisible by 100. We have ninety-nine consecutive numbers here, and each one of them is composite. Thus we have found a gap of at least ninety-nine composite numbers between consecutive primes.

Of course, if we started with say, 1,000!, a similar chain of reasoning would lead us to conclude that we had found a string of at least 999 composite numbers. Proceeding in this manner, we can find gaps as large as we like between consecutive primes. This result is extremely surprising and indeed counterintuitive.

The prime numbers have been in the news in recent years because of a movie based on a man who becomes obsessed with the number 23. Of course the number 23 consists of the first two primes concatenated together. Each cell in a human being normally contains 23 pairs of chromosomes, and $23 = 5 + 7 + 11$. The numbers in that equation are the first five primes in order. Also $23 = 3 \cdot 5 + 3 + 5$.

The twenty-third letter of the alphabet is *W*, which has two points pointing down and three points pointing up. Just above the letter *W* on a Qwerty keyboard are the digits 2 and 3. If you calculate 23! ($23 \cdot 22 \cdot 21 \cdot \ldots \cdot 3 \cdot 2 \cdot 1$), you will find that it contains twenty-three digits. The only prime number with this property is 23.

It has been claimed that the late mathematical genius Professor John Nash (1928–2015) was obsessed with the number 23. He claimed it was his favorite prime number. Curiously, the number 23 or its digits cropped up in various places in Nash's life. Here are a few examples. John Nash joined the faculty of the Massachusetts Institute of Technology (MIT) in 1951, when he was 23 years old. He met his future wife, Alicia Lardé (1933–2015) in this year also. Nash published 23 scientific papers in his lifetime. He shared the Nobel Memorial Prize in Economic Sciences with the German economist Reinhard Selton (1930–) and the Hungarian-American economist John Harsanyi (1920–) in 1994. The digits of 1994 sum to 23.

John and his wife, Alicia, were tragically killed in a car crash on the New Jersey Turnpike when they took a ride in a taxi on their way home from Norway on Saturday, May 23, 2015. (John had been to Norway to receive the Abel Prize in mathematics.) That date is the 23rd day of the $(2 + 3)$ month. The year, 15, equals $23 - 2^3$. The crash occurred near Interchange 8A. Note that 8 equals 2^3. The taxi driver, Tarek Girgis, survived the crash. He was aged 46 $(23 + 23)$ when the crash happened. The crash occurred at 4:30 p.m. and John and Alicia were pronounced dead at 5:18 pm. Note that 5 equals $2 + 3$ and that 18 equals $23 - 2 - 3$. John Nash was 86 years old when he was killed. There are 23 prime numbers less than 86.

Here's a little curiosity I discovered. Use the following number code for the letters of the alphabet: A = 1, B = 2, C = 3, and so on. Now add up the values of the letters in the word ODD, ODD, and EVEN. The following equation is true in the sense that the sum of two odd numbers equals an even number, but it is also true in the sense that the value of the letters in the words ODD + ODD equals the value of the letters in the word EVEN.

ODD + ODD = EVEN
23 + 23 = 46

Here's another fun curiosity. Consider the prime number 998,443. If you drop one digit at a time from the left-hand side of this number, you obtain a new prime after each digit is dropped. Thus, 98,443, 8,443, 443, 43, and 3 are all primes. This number, 998,443, is called a *left-truncatable prime.* It is the largest left-truncatable prime less than one million. There are 4,260 left-truncatable primes that we are aware of so far. The largest known left-truncatable prime is 357,686,312,646,216,567,629,137.

On the other hand, there are 83 known *right-truncatable primes*, of which the largest is 73,939,133. The largest right-truncatable prime less than one million is 739,399. Thus 73,939, 7,393, 739, 73, and 7 are all primes.

There are just fifteen known prime numbers that are both left-truncatable and right-truncatable. These are: 2, 3, 5, 7, 23, 37, 53, 73, 313, 317, 373, 797, 3,137, 3,797, and 739,397.

We mentioned that there are 4,260 left-truncatable primes and 83 right-truncatable primes. Interestingly, these two numbers are related as follows. Raise the digits of the number 4,260 to the power of 4 and sum the results (i.e., $4^4 + 2^4 + 6^4 + 0^4$). This equals 1,568. Now partition that number into the two numbers 15 and 68. Sum those two numbers to obtain 83.

The number 4,260 has a curious property: 4,260 + 1; $2 \cdot 4{,}260 + 1$; $3 \cdot 4{,}260 + 1$ and $4 \cdot 4{,}260 + 1$ are all prime numbers. The number 83 is curious also because $(83 - 1)/2 = 41$, which is prime, and $(83 \cdot 2) + 1 = 167$, which is also prime. The number 83 is also the sum of the first three primes that end in 1: 83 = 11 + 31 + 41.

The primes have fascinated me since I was a young child. (I have noted that even the word PRIME has a prime number of letters in its name. Using the alphabet code above, the sum of the letters in the word PRIME is 61, which is also prime. The word COMPOSITE [which means nonprime] has a composite number of letters in its name. Using the same alphabet code, the letters of the word COMPOSITE sum to 115, which is also composite.)

In my opinion, the distribution of the primes is one of the great mysteries of the world.

Your friend,
Dr. Cong

That was the e-mail I received from Dr. Cong.

Before I sign off, I will take this opportunity to give you a very brief biography of Ming Cong.

I was "introduced" to Dr. Cong by a mutual acquaintance a few months ago. Since then I have carried out some research on this rather strange and eccentric man. This is what I have discovered thus far.

Ming Cong was born in China on Tuesday, January 2, 1934 (the Gregorian calendar) at four minutes past 6:00 a.m. He weighed exactly eight pounds at birth. His father was a civil engineer and his mother held a PhD in mathematics. Apparently Ming was just four years old when he stunned his parents by telling them that he had a natural liking for numbers because he was born on 1/2/34 at 56 minutes to 7, he weighed exactly 8 pounds at birth, and he was born on the $\sqrt{9}$ day of the week.

His father died when Ming was six years old. His mother soon returned to the United States and raised her only child in New York.

I discovered Ming had gone to college to study mathematics and physics in his youth. It was around this time that his mother died. Young Ming was heartbroken, but to his credit continued his college education.

However, during a short break in his third year at college, he was involved in an accident while skiing with friends in Europe. Dr. Cong suffered severe injuries to his head and upper body and was never the same man again. He apparently lost all interest in what had promised to be a glittering career in mathematical research.

Instead he left his home and traveled with a carnival throughout the United States. Because he had been to college, Ming earned the nickname of "Doctor" from his work colleagues in the carnival, and soon he came to be widely known as Dr. Cong. The crafty numerologist appeared to like the nickname and began to enjoy the status that was bestowed upon him as a consequence of his new title. He never again referred to himself as Ming Cong but signed all letters and documents as "Dr. Cong."

Cong was known to give unusual and bizarre pieces of information to his colleagues in the carnival. For example, when Hawaii became the last state to join the union in 1959, Dr. Cong was able to tell them that Hawaii joined the USA 177 + 6 (183) years after 1776. He reminded one elderly lady in 1976 when Americans were celebrating the 200th anniversary of Independence that July 4, 1776, fell on the (17 + 76 + 17 + 76) day of the year. Dr. Cong told some friends that using the alphabet code A = 1, B = 2, C = 3, and so on, the letters in the phrase "PRESIDENT K IS BORN" sum to 191 + 7. He would then remind his listeners that President Kennedy was born in 1917.

On one occasion, Dr. Cong pointed out to a group of teenagers visiting the carnival that (using the earlier code) the letters of the phrase "THE NEW WORLD" sum to 149 – 2 (i.e., 147), bringing the year America was discovered, 1492, to mind. On another occasion, Dr. Cong pointed out to some patrons of the carnival that the letters of the phrase "THE WORLD WAR ENDS" sum to 194 – 5 (i.e., 189). Of course, the Second World War ended in 1945. Dr. Cong also once pointed out to a high-ranking NASA employee visiting the carnival that it was curious that the first Apollo flight that landed two men on the moon in 1969 should be numbered 11 because 11 equals $\sqrt{196} - \sqrt{9}$. The NASA employee was said to be dumbfounded by the curious fact.

Dr. Cong surprised colleagues when he mentioned a number curiosity about the Second World War at a Christmas party in 1945. That war began, Cong said, on September 1, 1939, when Germany invaded Poland. It officially ended on September 2, 1945. The war started in the year $4 + 5 + 6 + 7 + 8 + 9$ and ended in the year $1 + 2 + 3 + 4 + 5 + 6 + 7 + 8 + 9$. Cong said it was appropriate that the Second World War ended in '45 on the 245th day of the year.

Dr. Cong sometimes performed excellent card tricks at the carnival. An acquaintance of mine who witnessed some of these card tricks said that Cong was a master magician. He once pointed out to my acquaintance that if you count the letters in the thirteen-card values ace, two, three, four, five, six, seven, eight, nine, ten, jack, queen, and king, you will obtain 52, which is the number of cards in a standard deck.

During these magic sessions, he often dispatched many other odd facts to his listeners. For example, he told his audience on one occasion that (using the earlier alphabet code) if you add up the value of the letters in the number "TWO HUNDRED AND FIFTY ONE," you will obtain 251.

Dr. Cong began reading Martin Gardner's (1914–2010) monthly articles titled "Mathematical Games" in *Scientific American* from the very beginning of the series. Cong recently told me that this helped him keep up with what was happening in mathematics while he was on the road as a "carny."

Dr. Cong once mentioned to a mathematician visiting the carnival that Martin Gardner's first monthly article appeared in *Scientific American* in January 1957. Gardner often mentioned that one of his favorite numbers was pi (which we will meet later in this book). Pi equals 3.14159265 . . . . The decimal expansion is never-ending. Dr. Cong pointed out to this mathematician that Gardner was exactly 15,413 days old on January 1, 1957. That number, Cong said, contains the first five digits of pi. The mathematician, I have been told, almost fainted with shock at this news.

Dr. Cong was asked recently by a young mathematics student if he had any curiosities about the numbers 2013, 2014, and 2015. The old showman showed just how good he is with numbers when he almost immediately pointed out that 2013, 2014, and 2015 are each the product of three distinct primes. In other words, 2013 equals 3 times 11 times 61; 2014 equals 2 times 19 times 53; and 2015 equals 5 times 13 times 31.

Although Cong has clocked up the years, he has lost none of his old magic. Just a few weeks ago, he informed me by e-mail that the year 2016 would be the first of its kind since 1953. He pointed out that 2016 equals the sum of the numbers from 1 to 63. (The number 1953 equals the sum of the numbers from 1 to 62.) He also mentioned that 2016 equals $3^3 + 4^3 + 5^3 + 6^3 + 7^3 + 8^3 + 9^3$. He also gave me a neat fact about the capital city of my native country, Ireland. He said that using the earlier alphabet code the sum of the letters in the words "CAPITAL" and "DUBLIN" are both 62.

As the months turned into years, Dr. Cong turned his mathematical mind to various pursuits. One of these was professional gambling. Dr. Cong was said to be highly successful in this field. So much so that he has apparently just left the carnival scene in the last year or two and is reported to be now quite a wealthy man. He keeps to himself and is generally considered by his neighbors to be a nerd.

When I first spoke with him on the telephone, he appeared to genuinely believe in the power of numerology. I found this to be rather strange, but I did not comment on it. I had the pleasure of meeting him last month. I asked him if he always wanted to be a mathematician. He replied that when he was seven he went to see a comedian performing in his school play. He initially thought that this might be a good career to follow but soon realized he was afraid people might laugh at him. So he dropped that idea. Dr. Cong went on to tell me that, in his opinion, numbers have a life and a mind of their own! Of course, I could not, under any circumstances, agree with his opinion. So we agreed to disagree on this point.

Dr. Cong does, however, appear to have a mathematical mind. He is excellent at spotting primes, and he comes up with some really unusual properties of numbers. Dr. Cong is also quite adept at spotting the odd coincidence or curiosity.

I discovered only recently that his wife was killed in an automobile accident many years ago. He has an adult daughter, named Chan. She apparently also knows a thing or two about numbers. I think readers of this book would enjoy their company.

CHAPTER 3

SOME WORDS ON PYTHAGOREAN TRIPLES

The Pythagorean theorem is a theorem well known around the world to teenagers and adults alike. It has often been described by mathematicians as a very beautiful theorem. Therefore, before we explore this famous theorem and how it relates to primes, I think a few words about the nature of beauty may be appropriate.

For centuries, philosophers have debated the nature of beauty. Is beauty, as the old proverb states, merely in the eye of the beholder? Is it a subjective thing? Or is beauty an objective quality? Is a thing beautiful in itself, whether or not I or others think it is beautiful?

Most people—but certainly not all—feel intense pleasure when looking at a dramatic sunset on a summer's evening. Usually, the observer notices with delight the colors changing in the evening sky as the sun starts to get lower and lower, until eventually the western evening sky is filled with splashes of orange, pink, and red.

The feeling of joy and wonder is intensified if the observer realizes that the sun is a star 93 million miles from Earth, consisting of 10^{57} atoms.[1]

If that same observer contemplates the fact that she is made up of about $7 \cdot 10^{27}$ atoms, she may begin to realize that she, like the sun, is also a collection of atoms.[2] The thought may strike her: *A collection of atoms contemplating atoms tens of millions of miles away! How strange! How marvelous! To think that collections of atoms can write books about atoms! Who would have thought of it! What a wonderful achievement by nature!*

The philosopher Augustine of Hippo (354–430 BCE) asked this question[3]: Is a thing beautiful because it gives delight, or does it give delight because it is beautiful? He believed that the second proposition is the correct one. To this day, the question still reveals differences of opinion among philosophers.

If beauty exists in the eye of the beholder, does it have any reality outside that individual's mind? If it does not, then when we say something is beautiful, that "beauty" is merely reduced to the expression of our personal opinion. We have not given any *objective* information concerning the object in question to those we are communicating with. In fact, we cannot give any *objective* information on the relevant object. This is an unacceptable belief to have in the eyes of many philosophers.

On the other hand, the belief that something is beautiful in itself, whatever observers may or may not believe, leads us onto shaky ground also. If beauty in itself exists independent of perceivers, then that would imply that if there were no intelligent beings on this planet, beautiful sunsets, for example, would continue to exist. The question then arises, How could beautiful sunsets exist in any real sense if there are no intelligent observers to witness them and to describe them as being beautiful? This viewpoint is also unsatisfactory for many philosophers.

A somewhat similar philosophical problem arises over the question of beauty in mathematics. Is it correct to say that a mathematical theorem is beautiful? Is the theorem beautiful in itself? Or are we merely expressing our point of view when we utter such words?

Pure mathematics is the study of mathematic for its own sake. Applied mathematics is the study of mathematics so that the results obtained can be applied in the physical world. The applied mathematician is delighted when his results are used in computer engineering, medicine, television, and a host of other technologies. Our affluent and comfortable way of life is possible only because applied mathematicians have made good use of the knowledge of mathematics that we have accumulated over the years.

On the other hand, the pure mathematician studies mathematics because she is intensely interested in mathematics for its own sake. She is not interested or concerned whether the results obtained from her efforts might be useful in the physical world. She studies mathematics because she believes that it is an extremely interesting subject that reveals a hidden, beautiful order behind mathematical objects. The pure mathematician may even state that she studies mathematics because she finds many of the mathematical results "beautiful."

Consider the following scene which is—most likely—played out in some part of the world every day. The pure mathematician (or recreational mathematician) sits at her desk with the intention of beginning to explore some mathematical topic that she may know very little about. Say that topic is the Pythagorean theorem. She wants to inquire particularly into primitive Pythagorean triangles. These are right triangles whose three sides are all integers.

The pure mathematician will almost certainly know the basics of the object of study: that the Pythagorean theorem states that in any right triangle the square on the longest side of the triangle (the longest side of a triangle is called the *hypotenuse*) is equal in area to the sum of the squares on the two other sides (the other two sides of a triangle are called *legs*).

Before she begins her inquiries at all, the pure mathematician has a deep belief that she will find beautiful connections between the integers that are associated with right triangles. She does not know yet what these connections are, or to what degree she will consider them beautiful. But she is convinced that these connections are real and that they already exist. She has no doubt that these connections are "out there" and are waiting to be "discovered." She certainly does not believe that she is going to "invent" these connections.

The pure mathematician has no idea how trivial or deep these connections that she may discover are going to be. They may be of no practical use whatsoever in the physical world. That does not concern her. She has a deep belief, however, that whatever relations she finds will be beautiful. She does not have any particular work strategy to produce a particular result. Far from it! The pure mathematician performs her inquiries in an open manner by following her nose, so to speak. She believes that whatever results she obtains were there all the time and that she is forced to accept those results, whatever she may think of them! In other words, she believes that mathematical reality will impose its truths upon her and indeed upon any other inquirer.

In a nutshell, the pure mathematician believes that she has no choice but to accept the mathematical results that mathematical reality forces upon her. She is convinced that those mathematical theorems that she may discover have always been true, long before she entered the world, and indeed, long after she will leave the world. If she thinks about the issue sufficiently, she will probably come to the conclusion that mathematical theorems are eternal and timeless: they have always been true and always will be.

This point of view that mathematics is beautiful and eternal and that it has an independent existence from human minds is sometimes used as an argument by mathematicians who hold the opposite point of view. They argue that mathematicians who believe that mathematical reality exists "out there" also believe in the existence of God.

This may be true in particular cases, but is certainly not always the case. Many mathematicians who believe in the external reality of mathematics do not believe in any creator. To cite just one example, Bertrand Russell, the great British philosopher and mathematician, fervently believed in mathematical beauty and that mathematics exists outside our minds, but he most certainly did not believe in the existence of any god.[4]

If mathematical reality exists outside of our minds, then the only thing we can conclude from that is that mathematical reality exists outside of our minds! It does not prove the existence of a god or gods. In fact, a number of mathematicians and philosophers strongly make the point that the universe exists as a consequence of the external and eternal nature of mathematical reality.

It is well known to philosophers that theists often use an argument that because great works of art have been created by man that this somehow indicates that there must be a god overlooking the works of man. Richard Dawkins makes reference to this argument in his book *The God Delusion*.[5] Dawkins states that on numerous occasions over the years people have asked him, how does one account for the beauty of Shakespeare's sonnets, if there is no God? Dawkins's response is to point out to the inquirer that Shakespeare's sonnets are indeed beautiful, but that that fact does not prove the existence of God; it proves the existence of Shakespeare!

When students of recreational mathematics first encounter the Pythagorean theorem and primitive Pythagorean triples, they also frequently ask themselves if these relations between the integers found in primitive Pythagorean triples have a reality of their own. They certainly get the feeling that this theorem, which has been known to humanity for thousands of years, has always been somehow true. Whether this is in fact the case or whether this is an illusion or trick played on each of us by our brain is, of course, hotly debated to this day.

What is this theorem that has been described by numerous mathematicians over the years as one of the most widely known in the whole of mathematics? The Pythagorean theorem is simple to state, yet it is widely considered among mathematicians to be a beautiful theorem. Even adults who have very little interest in mathematics will often state in conversation that one of the theorems they remember most from their school days is the Pythagorean theorem. The theorem, which concerns a relation in Euclidean geometry, actually predates Pythagoras and is believed by many to be nearly four thousand years old.[6]

The Pythagorean theorem refers to right-angled triangles. These are triangles that contain a 90-degree angle. The Pythagorean theorem states that in any right-angled triangle, or right triangle, no matter how big or small it may be, the square on the hypotenuse of the triangle is equal to the sums of the two squares on the legs of the triangle.

It is customary to write the lengths of each of the two legs of any right triangle as a and b, and the length of the hypotenuse as c. Using this notation, the Pythagorean theorem may be written as follows: $a^2 + b^2 = c^2$.

The theorem is so simple to state that a child familiar with squares can understand what the theorem asserts. Yet this theorem is widely used across different branches of mathematics. Before I give a simple proof of the Pythagorean theorem, I will remind nonmathematical readers that the area of any right triangle is equal to the product of the length of the two legs divided by 2.

Here's a simple proof of the Pythagorean theorem. Consider figure 3.1.

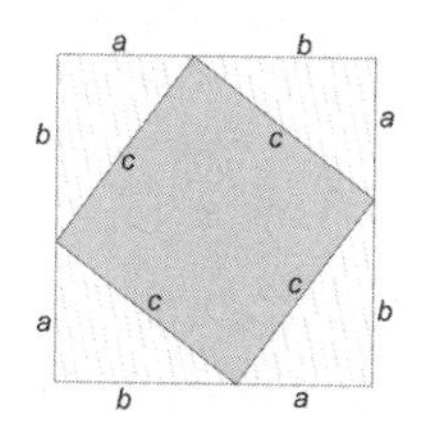

Figure 3.1

We arrange any four right-angled triangles of equal size, with legs lengths a and b and a hypotenuse of length c, as shown in figure 3.1. The areas of each of these triangles are shown with light, broken lines running through them.

Consider the larger square. The length of the side of the larger square is $a + b$. The smaller square in the center of the larger square is actually the square built on any one of the hypotenuses of the right triangles. Thus the area of the smaller square can be designated c^2.

Since the length of the side of the larger square is $a + b$, its area is $(a + b)^2$. This equals $a^2 + 2ab + b^2$. This area is equivalent to the total areas of the four right triangles, plus the area of the center square, which we have designated c^2.

The areas of each of the four right triangles are obtained by the usual formula for calculating the area of any right triangle. This formula is to multiply the lengths of the two legs and then divide that result by 2. The result is the area of the triangle. Thus the area of each of the four right triangles is $\frac{1}{2}ab$. Since there are four such right triangles in figure 3.1, the total area of the four right triangles is $\frac{1}{2}ab \cdot 4$. This equals $2ab$.

This allows us to write:

Total area of large square whose side length is $a + b$ = *Area of four right triangles plus* c^2.

This equals $a^2 + 2ab + b^2 = 2ab + c^2$.

Subtract $2ab$ from each side of the equation to obtain $a^2 + b^2 = c^2$.

This result proves the Pythagorean theorem.

The converse of the theorem can be proved also. If three sides of a triangle a, b, c are such that $a^2 + b^2 = c^2$, then that triangle is a right triangle.

Of course, the values of a, b, and c do not have to be integers. If a equals, say, 123.45 and b equals 67.89, we can write:

$123.45^2 + 67.89^2$ = square of some quantity; say it is x^2

Thus

$15{,}239.902 + 4{,}609.052 = x^2$

or

$19{,}848.954 = x^2$

Therefore,

$$\sqrt{19{,}848.954} = x$$

$$\textit{Thus } 140.886 = x.$$

Numerous articles and books have been published over the years giving various proofs of this famous theorem. There have been well over three hundred and fifty proofs of the theorem discovered. In fact, one such proof was given by the United States president James Garfield in 1876, before he became president.[7]

When the three sides of a right triangle are all integers, those three integers combined are known as a *Pythagorean triple*. For example, the right triangle with the smallest possible integers is the 3, 4, 5 right triangle. (The reader will notice that $3^2 + 4^2 = 5^2$.) Thus the three integers 3, 4, and 5 are collectively known as a Pythagorean triple. The second smallest Pythagorean triple is 5, 12, and 13 because $5^2 + 12^2 = 13^2$.

When a, b, c have no common divisors, the triple is known as a *primitive Pythagorean triple*. For example, the Pythagorean triple 6, 8, and 10 is a Pythagorean triple because $6^2 + 8^2 = 10^2$. However, these three numbers have a common divisor: They are all divisible by 2. If we divide each number in the triple by 2, we obtain 3, 4, and 5. This is also a Pythagorean triple because $3^2 + 4^2 = 5^2$. But the reader will find that the three integers 3, 4, and 5 have no common divisor. Therefore, that triple cannot be reduced down to smaller integers. When a Pythagorean triple cannot be reduced any further, that Pythagorean triple is known as a *primitive Pythagorean triple*, or PPT, for short. Therefore, the triple 3, 4, 5 is a primitive Pythagorean triple or PPT.

Primitive Pythagorean triples in which the hypotenuse is greater than the longer leg by one unit have a special appeal to many mathematicians because they exhibit some beautiful, simple patterns, which we will discuss in a moment. Here we shall describe such primitive Pythagorean triples as *special primitive Pythagorean triples*. The first ten smallest *special primitive Pythagorean triples* are:

$$\begin{aligned}
3^2 + 4^2 &= 5^2 \\
5^2 + 12^2 &= 13^2 \\
7^2 + 24^2 &= 25^2 \\
9^2 + 40^2 &= 41^2 \\
11^2 + 60^2 &= 61^2 \\
13^2 + 84^2 &= 85^2 \\
15^2 + 112^2 &= 113^2 \\
17^2 + 144^2 &= 145^2 \\
19^2 + 180^2 &= 181^2 \\
21^2 + 220^2 &= 221^2
\end{aligned}$$

Figure 3.2

Consider the primitive Pythagorean triples as shown in figure 3.2. First, note that all the left-hand integers in the series of PPTs are the odd numbers beginning with 3. The series runs 3, 5, 7, 9, 11, 13, 15, 17, 19, 21, and so on.

The series of numbers immediately left of the equation signs in figure 3.2 is 4, 12, 24, 40, 60, 84, 112, 180, 220, and so on. These are the longer legs in the special PPTs. Consider the following pattern involving these numbers:

$$
\begin{aligned}
4 \cdot 1 &= 4 \\
4 \cdot (1 + 2) &= 12 \\
4 \cdot (1 + 2 + 3) &= 24 \\
4 \cdot (1 + 2 + 3 + 4) &= 40 \\
4 \cdot (1 + 2 + 3 + 4 + 5) &= 60 \\
4 \cdot (1 + 2 + 3 + 4 + 5 + 6) &= 84 \\
4 \cdot (1 + 2 + 3 + 4 + 5 + 6 + 7) &= 112 \\
4 \cdot (1 + 2 + 3 + 4 + 5 + 6 + 7 + 8) &= 144 \\
4 \cdot (1 + 2 + 3 + 4 + 5 + 6 + 7 + 8 + 9) &= 180 \\
4 \cdot (1 + 2 + 3 + 4 + 5 + 6 + 7 + 8 + 9 + 10) &= 220
\end{aligned}
$$

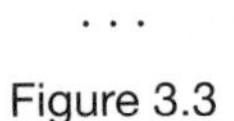

Figure 3.3

Note that the sum of the two legs in special PPTs forms the following pattern:

$$
\begin{aligned}
3 + 4 &= 7 &&= 2 \cdot 2^2 - 1 \\
5 + 12 &= 17 &&= 2 \cdot 3^2 - 1 \\
7 + 24 &= 31 &&= 2 \cdot 4^2 - 1 \\
9 + 40 &= 49 &&= 2 \cdot 5^2 - 1 \\
11 + 60 &= 71 &&= 2 \cdot 6^2 - 1 \\
13 + 84 &= 97 &&= 2 \cdot 7^2 - 1
\end{aligned}
$$

$$\ldots$$

Figure 3.4

There are an infinite number of special PPTs whose hypotenuses are one unit longer than the longer leg. These special PPTs are easily found by using the following formula, where m and n are integers, one even and one odd, and m and n are both co-prime; that is, m and n have no common divisor, except 1: $a = m^2 - n^2$; $b = 2mn$; $c = m^2 + n^2$. The formula produces all special primitive Pythagorean triples. If m and n are both even, or both odd, the formula still produces a Pythagorean triangle, but not a *primitive* Pythagorean triangle. The formula produces special PPTs if, and only if, the generating numbers m and n, where m is greater than n, differ by 1. Figure 3.5 shows how these generating numbers, m and n, produce special *PPTs*.

	a	b	c	$\rightarrow$	a	b	c
$m\ n$	$m^2 - n^2$	$2mn$;	$m^2 + n^2$				
2 1	$2^2 - 1^2$	$2 \cdot 2 \cdot 1$	$2^2 + 1^2$	$\rightarrow$	3	4	5
3 2	$3^2 - 2^2$	$2 \cdot 3 \cdot 2$	$3^2 + 2^2$	$\rightarrow$	5	12	13
4 3	$4^2 - 3^2$	$2 \cdot 4 \cdot 3$	$4^2 + 3^2$	$\rightarrow$	7	24	25
5 4	$5^2 - 4^2$	$2 \cdot 5 \cdot 4$	$5^2 + 4^2$	$\rightarrow$	9	40	41
6 5	$6^2 - 5^2$	$2 \cdot 6 \cdot 5$	$6^2 + 5^2$	$\rightarrow$	11	60	61
				...			

Figure 3.5

If we let A equal the odd leg in such triples, the hypotenuse equals $(A^2 + 1)/2$.

If we multiply each of the three numbers in any primitive Pythagorean triple by any integer, the result will be a Pythagorean triple, but not, of course, a primitive Pythagorean triple. For example, if we multiply the 3, 4, 5 triple by 2, we obtain the 6, 8, 10 triple. Thus $6^2 + 8^2 = 10^2$. If we multiply the 3, 4, 5 triple by 3, we obtain the 9, 12, 15 triple. Thus $9^2 + 12^2 = 15^2$. If we multiply the triple 11, 60, 61 by say, 23, we obtain the triple 253, 1,380, and 1,403. We find that $253^2 + 1{,}380^2 = 1{,}403^2$. In this way we can find an infinite number of Pythagorean triples.

Pythagorean triples have been studied by mathematicians down through the centuries, and consequently an enormous amount of literature exists on the subject. A multitude of proofs of various identities concerning Pythagorean triples have been found scattered through this vast literature. We will give some results here without the proofs.

In any Pythagorean triple, one side is always divisible by 3, one side is always divisible by 4, and one side is always divisible by 5.[8] Occasionally, one side will be divisible by two of these divisors. For example, in the 5, 12, 13 triple, the hypotenuse is not divisible by 3, 4, or 5. However, the shortest side is divisible by 5, and the other leg is divisible by both 3 *and* 4.

Even less occasionally, all three divisors can appear in one leg of the Pythagorean triple, as in the 11, 60, 61, triple. Here the leg that is 11 in length and the hypotenuse (which is 61 in length) is not divisible by 3, 4, or 5. But the leg that is 60 in length is divisible by 3, 4, *and* 5. As a consequence of the above rule, the area of every Pythagorean triple is a multiple of 6, and the product of the three sides of every Pythagorean triple is a multiple of 60.

Every integer, except 1, 2, or 4, can be the shortest side of a Pythagorean triangle.

In any Pythagorean triple, the sum of the three sides, $a + b + c$, will divide ab. For example, in the Pythagorean triple 3, 4, and 5, the sum of the three numbers is 12 and the product of the two legs is 3 times 4, or 12. We see that that the sum of the three sides divides into the product of the lengths of the two legs. In the 5, 12, 13 Pythagorean triple, the sum of the three sides is 30 and the product of the two legs is 60. We see that 30 divides into 60. And so on.

If a circle is inscribed inside a special primitive Pythagorean triangle, the radius of that circle is always an integer.[9] (See figure 3.6.)

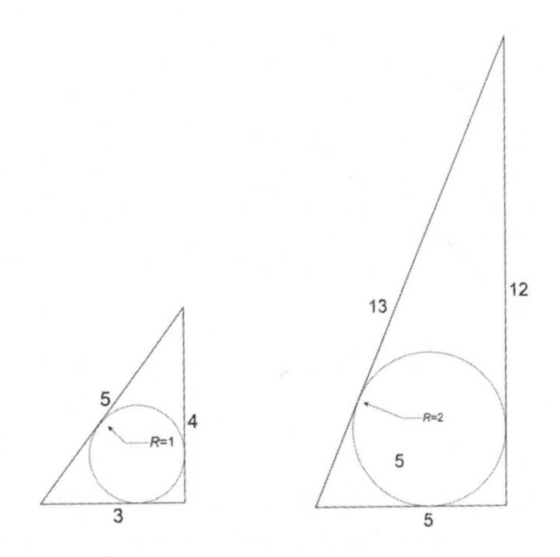

Figure 3.6

For the 3, 4, 5 triangle, the radius is 1; for the 5, 12, 13 triangle, the radius is 2. Figure 3.7 gives the radii of the first five special primitive Pythagorean triangles.

Special Primitive Pythagorean Triangle	Radius of Inscribed Circle
3, 4, 5	1
5, 12, 13	2
7, 24, 25	3
9, 40, 41	4
11, 60, 61	5
. . .	. . .

Figure 3.7

Note that in each special PPT, the sum of the hypotenuse and the radius of the inscribed circle are equal to half the perimeter of the relevant triple. For example, in the smallest PPT, 5 plus 1 equals 6, and this is half the perimeter of the 3, 4, and 5 triple. In the 5, 12, 13 triple, the hypotenuse 13 plus the radius of the inscribed circle, 2, equals 15, which is half the perimeter of the 5, 12, 13 triple.

Some curious facts about primitive Pythagorean triples have been known for generations. For instance, the sum of the hypotenuse and the even leg is always the square of an odd number. In the 3, 4, 5 triple, for example, $5 + 4$ equals 9, which is 3^2. The sum of the hypotenuse and the odd leg is always twice a square. For instance, in the 7, 24, 25 triangle, $25 + 7$ is twice 4^2. It is impossible for a Pythagorean triple to exist in which its hypotenuse and one leg are the legs of another Pythagorean triple. The great French mathematician Pierre de Fermat (1601–1665) proved that it is impossible for the area of a Pythagorean triangle to equal a square number.

The numbers that appear as the lengths of the two legs in special primitive Pythagorean triples follow the following simple pattern. The left-most integer in each row in figure 3.8 is the short leg, and the integer just right of the equation sign is the long leg:

$$3 \cdot 1 + 1 = 4 \quad \rightarrow \quad 3, 4, 5$$
$$5 \cdot 2 + 2 = 12 \quad \rightarrow \quad 5, 12\ 13$$
$$7 \cdot 3 + 3 = 24 \quad \rightarrow \quad 7, 24, 25$$
$$9 \cdot 4 + 4 = 40 \quad \rightarrow \quad 9, 40, 41$$
$$11 \cdot 5 + 5 = 60 \quad \rightarrow \quad 11, 60, 61$$
$$13 \cdot 6 + 6 = 84 \quad \rightarrow \quad 13, 84, 85$$
$$\ldots$$

Figure 3.8

A perfect number is a number that equals the sum of all its divisors, except itself. For example, 6 is an even perfect number. The divisors of 6 are 1, 2, 3, and 6. The sum of its divisors (except 6) is $1 + 2 + 3$, which equals 6. The number 6 is the smallest perfect number. The second smallest perfect number is 28. The divisors of 28 are 1, 2, 4, 7, 14, and 28. The sum of the divisors of 28 (excluding 28) is 28. The first four even perfect numbers are 6, 28, 496, and 8,128. Because mathematicians are curious by nature, many have wondered if even perfect numbers can be the hypotenuse of a Pythagorean triangle. The answer is they cannot.[10]

No one knows if odd perfect numbers exist. But it is known that if they do, they will possess certain number properties. One of these properties is that they can be the hypotenuse of a Pythagorean triangle.[11]

The area of a Pythagorean triangle is equal to the product of the two legs, divided by 2. Thus the area of the 3, 4, 5 right triangle is (3 · 4)/2, which equals 6, while its perimeter equals 12. Thus the area divided by the perimeter equals 0.5. There are, however, two Pythagorean triangles in which the area divided by the perimeter equals 1. These are the 5, 12, 13 triangle and the 6, 8, 10 triangle. There are three Pythagorean triangles in which the area divided by the perimeter equals 2. These are the 9, 40, 41; 10, 24, 26; and 12, 16, 20 triangles. There are six Pythagorean triangles in which the area divided by the perimeter is 3. These are the 13, 84, 85; 14, 48, 50; 15, 36, 39; 16, 30, 34; 18, 24, 30; and 20, 21, 29 triangles. The number of different Pythagorean triples whose ratios of areas to perimeters form the series 1, 2, 3, 4, 5 . . . is: 2, 3, 6; 4, 6, 9; 6, 5, 10; 9, 6, 12; 6, 9, 18; . . . and so on.

An interesting and unexpected pattern emerges in the above series of numbers. Each of these numbers is one-half the numbers of divisors of $8n^2$. When n equals 1, $8n^2$ equals 8, which has 4 divisors: 1, 2, 4, and 8; hence the number 2 in the series above. When n equals 2, $8n^2$ equals 32, which has 6 divisors: 1, 2, 4, 8, 16, and 32; hence 3 in the above series. When n equals 3, $8n^2$ equals 72, which has 12 divisors: 1, 2, 3, 4, 6, 8, 9, 12, 18, 24, 36, and 72; hence 6 in the above series. And so on.

The simple and beautiful order that lies hidden behind special primitive Pythagorean triangles may be illustrated by inspecting their perimeters: 12, 30, 56, 90, 132, 182, 240, and so on. These perimeters equal 3 · 4; 5 · 6; 7 · 8; 9 · 10; 11 · 12; 13 · 14; 15 · 60 . . . . This hints that there appears to be a beautiful order embedded in the fabric of mathematics, and that often this order is hidden. We find that we have to scratch below the surface to uncover this order. Why this is so is a mystery.

The following is a little-known method (even to teachers of mathematics) of calculating the area of a special primitive Pythagorean triangle. Let S equal the semiperimeter of the triangle, and c its hypotenuse. (The perimeter of a triangle is the sum of the lengths of its three sides. Thus the semiperimeter is half of this sum.) Then $S(S-c)$ will equal the area. In the 3, 4, 5 triangle, for example, S equals 6 and c equals 5. The formula $S(S-c)$ then equals 6 · (6 – 5) or 6. The area of the 3, 4, 5 right triangle is 6. In the 5, 12, 13 right triangle,

S equals 15 and c equals 13. The formula $S(S - c)$ then equals $15 \cdot (2)$ or 30. The area of the 5, 12, 13 right triangle is 30. And so on.

The areas of special PPTs are usually obtained as follows:

Special PPT	Area
3, 4, 5	$(3 \cdot 4)/2 = 6$
5, 12, 13	$(5 \cdot 12)/2 = 30$
7, 24, 25	$(7 \cdot 24)/2 = 84$
9, 40, 41	$(9 \cdot 40/2 = 180$
11, 60, 61	$(11 \cdot 60)/2 = 330$
13, 84, 85	$(13 \cdot 84)/2 = 546$

. . .

Figure 3.9

These areas can be expressed by the following simple series:

$$1 \cdot 2^2 + 2 = 6$$
$$3 \cdot 3^2 + 3 = 30$$
$$5 \cdot 4^2 + 4 = 84$$
$$7 \cdot 5^2 + 5 = 180$$
$$9 \cdot 6^2 + 6 = 330$$
$$11 \cdot 7^2 + 7 = 546$$

. . .

Figure 3.10

Note the following pattern involving these areas:

$$6 = 1 + 2 + 3$$
$$6 + 30 = 1 + 2 + 3 \ldots + 7 + 8$$
$$6 + 30 + 84 = 1 + 2 + 3 + \ldots + 14 + 15$$
$$6 + 30 + 84 + 180 = 1 + 2 + 3 + \ldots + 23 + 24$$
$$6 + 30 + 84 + 180 + 330 = 1 + 2 + 3 + \ldots + 34 + 35$$
$$6 + 30 + 84 + 180 + 330 + 546 = 1 + 2 + 3 + \ldots + 47 + 48$$

. . .

Figure 3.11

(Note that the final digit in each row to be summed in figure 3.11 is 1 less than a square number.)

The problem of finding integer solutions to $a^n + b^n = c^n$, when n is greater than 2, has a long and distinguished history in the world of mathematics that has been well documented over the years. Using very advanced mathematical techniques, the English mathematician Andrew Wiles (1953–) and his former student Richard Taylor (1962–) found a proof in 1995 that there are no solutions in non-zero integers to $a^n + b^n = c^n$, when the exponent, n, is greater than 2.[12]

Although Wiles and Taylor have shown that it is impossible to have an equation such as $a^n + b^n = c^n$, where a, b, and c are integers, and n is an integer greater than 2, some close calls to exceptions to the conjecture are possible. For example, $6^3 + 8^3 = (6 \cdot 6 \cdot 6) + (8 \cdot 8 \cdot 8) = 728$. Now $9^3 = (9 \cdot 9 \cdot 9) = 729$. Thus $6^3 + 8^3$ is just 1 short of 9^3.

Or consider the result $71^3 + 138^3 = 144^3 - 1$. In other words, $71^3 = (71 \cdot 71 \cdot 71) = 357{,}911$ and $138^3 = (138 \cdot 138 \cdot 138) = 2{,}628{,}072$. We find that $357{,}911 + 2{,}628{,}072 = 2{,}985{,}983$. We find that $144^3 = (144 \cdot 144 \cdot 144) = 2{,}985{,}984$. So again we find that the sum of two cubes, in this case $71^3 + 138^3$, is just 1 short of another cube.

So many surprising results have been found in Pythagorean triangles and triples, it is difficult to establish how many of these are widely known among readers of books on recreational mathematics. Here is one result that I believe is not widely known. Consider the area of each relevant successive special PPT (e.g., consider the 3, 4, 5 triangle.). Multiply the area by the shortest leg. Subtract the result from the square of the hypotenuse. The answer will always be the difference between two successive cubes!

For example, consider the second smallest special PPT, which is the 5, 12, 13 right triangle. The area of this PPT is 30. (The reader may recall we always find the area by dividing the product of the sides by 2.) Multiplying this by 5 and subtracting from the square of the hypotenuse (in this case, 169) gives 19, and 19 equals $3^3 - 2^3$. Let's do the same thing on the 7, 24, 25 right triangle. The area of this triangle is 84. Multiply 84 by the shortest leg, which is 7. The result is 588. Subtract 588 from the square of the hypotenuse ($25^2 = 625$), and the answer is 37. We find that 37 equals $4^3 - 3^3$. Performing a similar procedure on the 9, 40, 41 triangle gives 61, and 61 equals $5^3 - 4^3$. And so on.

In 2009, Joshua Zucker (a mathematics teacher at Castilleja School, Palo Alto, California) found the following beautiful pattern, where the successive hypotenuses of special primitive Pythagorean triples, 5, 13, 25, 41, 61, 85 . . . , fall on a diagonal line running from northwest to southeast:

1	2	4	7	11	16	22	29	37	46
3	5	8	12	17	23	30	38	47	57
6	9	13	18	24	31	39	48	58	69
10	14	19	25	32	40	49	59	70	82
15	20	26	33	41	50	60	71	83	96
21	27	34	42	51	61	72	84	97	111
28	35	43	52	62	73	85	98	112	127
36	44	53	63	74	86	99	113	128	144
45	54	64	75	87	100	114	129	145	162
55	65	76	88	101	115	130	146	163	181
66	77	89	102	116	131	147	164	182	201

Figure 3.12

The Pythagorean theorem states that the square of the hypotenuse of a right triangle is equal to the sum of the squares on the other two sides. Why does the Pythagorean theorem only work on right triangles? Because it is a special case of a more general mathematical equation, known as the *law of cosines* represented by:

$$c^2 = a^2 + b^2 - 2ab \cos C.$$

The cosine of an angle is used in trigonometry. It can be used to calculate the third side of any triangle (not necessarily a right triangle) when we know the lengths of two sides and the angle between them.

The cosine is explained as follows: Suppose we have a right triangle that also contains a 60-degree angle. The third angle will then be 30 degrees, because the three angles in any triangle add up to 180 degrees. The cosine of the 60-degree angle is the ratio between the length of the hypotenuse and the length of the adjacent leg. If you use a scientific calculator to find the cosine of 60 degrees, the calculator will tell you it is 0.5. What this tells you is that the length of the adjacent leg to the hypotenuse is one-half the length of the hypotenuse in any right triangle that has a 60-degree angle. Therefore, if the hypotenuse is 1 unit long, the adjacent leg will be 0.5 units in length. If you use the calculator to obtain the cosine of, say, a 30-degree angle, the calculator will give back the answer: 0.86602540 . . . . This tells you that the ratio between the hypotenuse and the adjacent leg to the hypotenuse of the triangle is 0.86602540 . . . . Again, if the hypotenuse is 1 unit in length, the adjacent leg (in this case it will be the longer of the two legs) is 0.86602540 units in length.

We return now to the equation:

$$c^2 = a^2 + b^2 - 2ab \cos C.$$

We first note that zero multiplied by any number equals zero. Therefore, if we can find a value for c in degrees that will make the cos C equal to zero, then the expression $2ab \cos C$ will also equal zero and the above equation will then reduce to

$$c^2 = a^2 + b^2 - 0$$

or

$$a^2 + b^2 = c^2.$$

It so happens that the cosine of 90 degrees is equal to zero. So if we have a triangle that contains a 90-degree angle, the cosine of that 90-degree angle will equal zero. In other words, where a and b are the two legs and c is the hypotenuse in the right triangle, the angle C (which is the angle opposite the hypotenuse) will be the right angle in the triangle. The law of cosines states that in *any* triangle the square on the hypotenuse is equal to the sum of the squares on the two legs minus $2ab \cos C$. In a right triangle, the presence of the right angle makes $2ab \cos C$ equal zero. Thus, in a right triangle the term $2ab \cos C$ vanishes and we are left with $c^2 = a^2 + b^2$. This is the Pythagorean theorem.

The Pythagorean theorem can be generalized to include any geometrical figures constructed on each of the three sides of a right triangle. In other words, if one takes any regular figure and places it on the sides of a right triangle, then the area of the figure on the hypotenuse is equal to the sum of the areas of the two similar figures on the two shorter sides. For example, if two semicircles of radius 3 and 4 are constructed on the legs of a Pythagorean triangle, the semicircle constructed on the hypotenuse will have a radius of 5. The combined areas of the two semicircles on the legs of the triangle will then equal the area of the semicircle on the hypotenuse. See figure 3.13.

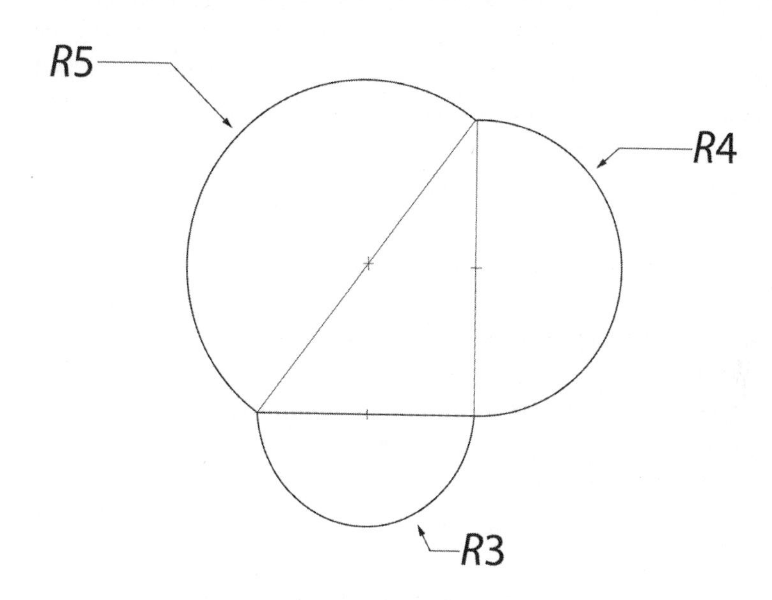

Figure 3.13

The area of a circle with radius r equals πr^2. Therefore the area of a semicircle is $\pi r^2/2$. In figure 3.13, the area of the smallest semicircle is $9\pi/2$ and the area of the second smallest semicircle is $16\pi/2$. The area of the semicircle on the hypotenuse is $25\pi/2$. Thus the combined areas of the two semicircles is $9\pi/2 + 16\pi/2$. This is equal to the area of the semicircle on the hypotenuse.

We know that the Pythagorean theorem holds only on plane or flat surfaces. It does not hold, for example, on the surface of a sphere or on a curved surface. Various forms of the equation, however, turn up in areas of mathematics or physics that apparently have nothing to do with right triangles. Sometimes these appearances of the Pythagorean triangle relation may at first seem mysterious to the casual observer, but in many cases we find on further investigation that there is nothing mysterious about the appearance.

For example, consider how the Pythagorean relation appears in the following mathematical puzzle. Suppose there are two large cylindrical tanks, equal in size. At the bottom of one tank, call it Tank A, a hole five inches in radius is cut. That hole is filled with a suitable-sized stopper. At the bottom of the other tank, call it Tank B, two holes very close to each other are cut. One of these holes is three inches in radius, and the second hole is four inches in radius. Both of these holes are also filled with a suitable-sized stopper.

Both tanks are filled with an equal amount of water. The three stoppers are removed at the same instant from the bottom of both tanks. Which tank will empty first? The answer is that both tanks will empty at the same instant. The two holes, three inches and four inches in radius, in Tank B, have a combined area equal to the hole five inches in radius in Tank A. So the amount of water will that will flow from Tank B will equal the amount that will flow from Tank A. Once the radii of the holes are equal to a Pythagorean triple (with the hypotenuse radius in one tank and the two leg radii in another), the tanks will empty in an equal period of time. If the radii of the three holes are not part of a Pythagorean triple, the tanks will empty at different rates.

It seems strange at first sight that the Pythagorean relation crops up in the solution to this puzzle. However, there is nothing remarkable about this appearance. The Pythagorean relation in the solution of this puzzle is a consequence of the fact that the area of a five-inch-radius circle equals the combined areas of a three-inch-radius and a four-inch-radius circle.

The Pythagorean relation also makes an appearance when an object such as a stone is dropped from rest from the top of a cliff. The stone will drop approximately $16.076 \cdot 1$, or 16.076, feet in one second. In two seconds, it will drop $16.076 \cdot 4$, or 64.304, feet. In three seconds, it will fall $16.076 \cdot 9$, or 144.684, feet; in four seconds, it will fall $16.076 \cdot 16$, or 257.216, feet. In five seconds, it will fall $16.076 \cdot 25$, or approximately 401.9, feet.

Note that the stone will fall in five seconds the combined distances it will fall in three and four seconds. In other words, $144.684 + 257.216 = 401.9$.

This relation will always hold as long as the three time periods in seconds that we choose are three numbers that form a Pythagorean triple. For example, consider the 5, 12, 13 Pythagorean triple. As mentioned earlier, the stone will fall 401.9 feet in five seconds. It will fall 2,314.944 feet in twelve seconds; and 2,716.844 feet in thirteen seconds. Note that $401.9 + 2{,}314.944 = 2{,}716.844$.

Like the previous puzzle, there is nothing remarkable about this apparent link between the Pythagorean relation and a stone dropping from rest. It results from the fact that squared numbers appear in Pythagorean relations and squared numbers also appear in calculating distances when objects are dropped from rest near the surface of the earth.

These results are, of course, trivial, and are mere oddities that do not have any deep mathematical significance. However, such results have caused many mathematicians and physicists to ask if there is a deeper link between the Pythagorean theorem and the laws of nature. The answer to this question is yes, but the answer was not discovered until a young man, unknown to the scientific community, started asking some basic questions about the way the laws of the universe operate.

His name was Albert Einstein. Among other things, Einstein discovered that nothing in the universe could travel faster through *space-time* than the speed of light. Incredibly, this fundamental law of nature is intimately related to the Pythagorean theorem.[13]

Here's how.

Einstein's marvelous discovery concerning the relation between energy, mass, and momentum that refers to objects *at rest* (not moving) is: $E = mc^2$. Here, E represents energy in ergs. (Physicists use the word *erg* to describe a unit of energy and work equal to 10^{-7} joules.) The lowercase m in Einstein's equation equals the mass (in grams) of an object, and c equals the speed of light (in centimeters per second). The equation tells us that mass and energy are interchangeable. In other words, mass and energy are two different forms of the same thing.

Einstein discovered the more general equation that refers to objects that are *in motion* (i.e., not in a state of rest): $E^2 = (mc^2)^2 + (pc)^2$. Once again, E equals energy, m equals mass, and c equals the speed of light. The pc in the equation equals the momentum expressed as a fraction of the speed of light, which is c. (If an object is moving in a certain direction, it has *momentum* in that direction. The faster the object moves, the more momentum it has.)

Einstein's equations tell us that the velocity of an object, divided by the speed of light, equals the ratio of pc to E in the right triangle. In other words, the E in the right triangle actually equals the speed of light. It was Einstein who first proposed this relationship between energy, mass, and momentum, but the relationship has been confirmed numerous times in experiments since then.

Let's take a closer look at the full equation: $E^2 = (mc^2)^2 + (pc)^2$. One of the first things to strike one on gazing upon this fundamental equation of nature is that it is similar to the Pythagorean theorem relationship. There is a squared quantity on the left-hand side of the equation. That squared quantity is equal to the sum of the squares of two other quantities that appear on the right-hand side of the equation.

Since E in the equation equals the speed of light, E is a fixed constant. In other words, the value of E cannot change. The value of mc^2 cannot change either, because the amount of mass that mc^2 represents remains conserved.

One may draw a right triangle to represent these three quantities. (See figure 3.14.) Let the base of the triangle equal the momentum of an object, pc, and the vertical leg equal mc^2. Let the hypotenuse equal E, which actually represents the speed of light.

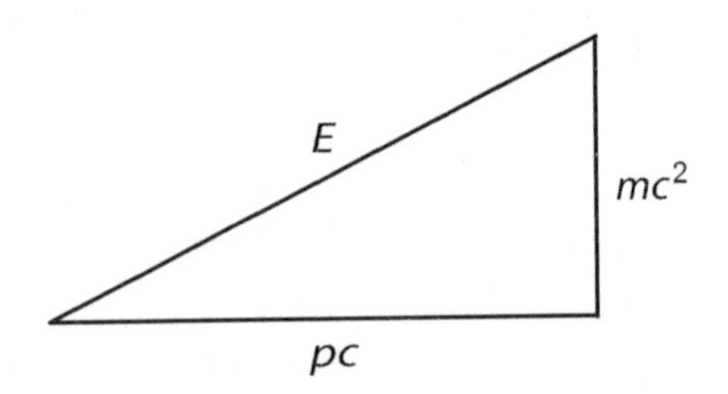

Figure 3.14

Because of the Pythagorean relationship involving mass, momentum, and energy it is not difficult to show that the momentum of an object with mass cannot ever equal the speed of light as the object moves through *space-time*. Why? Well, as the velocity of an object increases toward the speed of light, the base line in the triangle representing pc increases in length. The other leg of the triangle representing mc^2 stays the same length, because the mass that mc^2 represents stays constant.

As the velocity of the object continues to increase, the length pc gets much longer, but so also does the hypotenuse. (The hypotenuse has to get longer and longer—and must always stay longer than the longer leg—in order to retain the Pythagorean relationship between energy, mass, and momentum.) Therefore, as the length of the line pc increases in length (to represent the increasing speed of an object), the effect on the right triangle is that the length of pc approaches the value of the length of the hypotenuse. But the length pc will never *equal* the length of the hypotenuse, no matter how much speed the object is increasing by. Why? Because the length of the hypotenuse is always the longest line in any right triangle. Therefore, the length of the line pc, no matter what increase in velocity it may represent, can never attain the length of the hypotenuse. In other words, pc can never attain the speed of light.

Consequently, the speed of light is nature's speed barrier, and it can be justifiably argued that this is so because the hypotenuse of a right triangle is *always* longer than either of its two legs.

Physicists, cosmologists, mathematicians, philosophers, and mystics have found it curious, indeed amazing, that a law of nature that is analogous to the Pythagorean relationship between the hypotenuse of a right triangle and its two sides operates at the most fundamental level in the workings of the universe.

Incidentally, the interested reader may be asking, how did Einstein derive the specific equation $E = mc^2$ from the more general equation $E^2 = (mc^2)^2 + (pc)^2$? Think of it this way. The first equation applies when mass is *at rest*. The second equation applies when mass is *in motion* and therefore has momentum, and pc is a measure of that momentum.

An object's velocity V is equal to the ratio of the speed of light times the object's momentum over energy. If pc is the momentum of an object, we can therefore write. $V = c \cdot (pc/E)$. If the momentum continues to increase, pc gets closer and closer to equaling the energy, E. Thus $V = c \cdot (0.999 \ldots /E)$. Thus the ratio pc/E gets closer and closer to being equal to 1, and therefore the velocity gets closer and closer to equaling the speed of light. But because of the little piece of mass represented by the short leg of the right triangle, the longer leg of the triangle representing momentum cannot equal the hypotenuse of the right triangle, representing energy. Thus the velocity of an object traveling through space-time cannot ever equal the speed of light.

However, we can also see from figure 3.14 that if the value of pc is continually decreasing, the line pc is getting shorter and shorter in the right-triangle relationship. This shortening of the line pc has the effect of shortening the hypotenuse, E, also. As the length pc shrinks toward zero, the length of the hypotenuse, E, is approaching the length of the leg, mc^2. When pc becomes zero, the length of the hypotenuse, E, coincides with the length of the leg, mc^2. In other words, when matter is at rest, the hypotenuse E equals the leg mc^2. This is Einstein's famous equation: $E = mc^2$.

One can also see from figure 3.14 that if an object has no mass, such as a photon of light, then the short leg of the right triangle collapses to zero length and the length of E then equals pc. This tells us that if a massless particle exists, it has to have the same velocity as the speed of light. This, of course, is the case. Massless particles travel through space-time at the speed of light.

When I had completed this chapter, I sent it to my good friend Dr. Cong, asking for any comments or information he might care to give concerning Pythagorean triples. Dr. Cong kindly sent back the following information.

Hi, Owen,

Thank you for sending your paper on the Pythagorean theorem and Pythagorean triples.

The algebra proof of the Pythagorean theorem you gave is simple to follow. But there is an even simpler way of illustrating the truth of the theorem.

Consider the two identical squares below. [See figure 3.15.] The length of the square on the left is obviously $a + b$. The length of the large square on the right is also $a + b$. Thus those two large squares are equal in area.

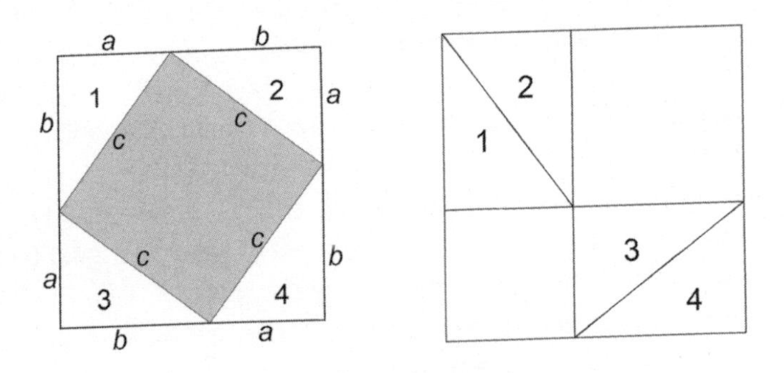

Figure 3.15

Consider the smaller internal square inside the large square on the left that has a side length equal to c. That smaller square is equal to the area of the larger square minus the areas of the four right triangles. Now consider the large square over on the right. There are two smaller squares inside the larger square. The areas of those two smaller squares are the squares on the two legs of the right triangle. The combined areas of those two smaller squares are equal to the area of the larger square minus the areas of the four right triangles. Therefore, the areas of those two smaller squares on the right must equal the area of the internal square with a side length of c on the left, thereby proving the Pythagorean theorem. This illustration not only proves the Pythagorean theorem *but also shows why it has to hold.*

The area of a triangle, or of any geometrical figure such as a square or a rectangle, is usually obtained as the *product* of two numbers. So it is quite pleasing to a mathematically minded individual to see the consecutive areas of primitive right triangles derived from the *sum* of consecutive numbers, as is shown here [see figure 3.16]:

$$1 + 2 + 3 = 6$$
$$4 + 5 + 6 + 7 + 8 = 30$$
$$9 + 10 + 11 + 12 + 13 + 14 + 15 = 84$$
$$16 + 17 + 18 + 19 + 20 + 21 + 22 + 23 + 24 = 180$$
$$25 + 26 + 27 + 28 + 29 + 30 + 31 + 32 + 33 + 34 + 35 = 330$$
$$\ldots$$

Figure 3.16

That pattern is easily remembered. Each row n commences with n^2. In other words, the first row begins with the number 1, which is the square of 1; the second row begins with 4, which is the square of 2; and so on. The number of terms being added to the left of the equation sign in each row is 3, 5, 7, 9, 11, and so on.

The following beautiful pattern, involving the area of special PPTs, is worth noting:

$$6 = 6 \cdot (1^2)$$
$$30 = 6 \cdot (1^2 + 2^2)$$
$$84 = 6 \cdot (1^2 + 2^2 + 3^2)$$
$$180 = 6 \cdot (1^2 + 2^2 + 3^2 + 4^2)$$
$$330 = 6 \cdot (1^2 + 2^2 + 3^2 + 4^2 + 5^2)$$
$$546 = 6 \cdot (1^2 + 2^2 + 3^2 + 4^2 + 5^2 + 6^2)$$
$$840 = 6 \cdot (1^2 + 2^2 + 3^2 + 4^2 + 5^2 + 6^2 + 7^2)$$
$$\ldots$$

Figure 3.17

The perimeters of the special PPTs also follow a beautiful and somewhat similar pattern:

Sides of Triangle	Perimeter of Triangle
3, 4, 5	2 (1 + 2 + 3)
5, 12, 13	2 (1 + 2 + 3 + 4 + 5)
7, 24, 25	2 (1 + 2 + 3 + . . . + 6 + 7)
9, 40, 41	2 (1+ 2 + 3 + . . . + 8 + 9)
11, 60, 61	2 (1 + 2 + 3 + . . . + 10 + 11)
13, 84, 85	2 (1 + 2 + 3 + . . . + 12 + 13)
15, 112, 113	2 (1 + 2 + 3 + . . . + 14 + 15)

Figure 3.18

I was surprised, Owen, that you did not include the following formula in your chapter:

$$2n^3 + 3n^2 + n.$$

When a positive integer is substituted for *n*, the formula produces the area of the *n*th special PPT. For example, when *n* equals 1, the formula equals $2 \cdot 1^3 + 3 \cdot 1^2 + 1 = 6$, which is the area of the smallest special PPT. When *n* equals 2, the formula equals $2 \cdot 2^3 + 3 \cdot 2^2 + 2 = 30$, which is the area of the second smallest special PPT. When *n* equals 3, the formula equals $2 \cdot 3^3 + 3 \cdot 3^2 + 3 = 84$, which is the area of the third smallest special PPT, and so on.

When I was a boy of nine years, I discovered the following little curiosity: The smallest Pythagorean triple involves the square numbers, 9, 16, and 25. The first two letters of the word *Pythagorean* are the 16th and 25th letters of the alphabet.

Finally, I will mention a curiosity concerning the 3, 4, 5 right triangle that is not well known: The number 12 is the perimeter of the 3, 4, 5 Pythagorean right triangle. Its area equals 6 square units, and $7 + 8 + 9$ equals twice the perimeter of the triangle. Note how the nine digits appear in order in the previous sentences!

Kind regards from your friend,

Dr. Cong

CHAPTER 4

THE MONTY HALL PROBLEM AND OTHER DECEPTIVE PUZZLES IN PROBABILITY THEORY

The following problem arose in a TV show in the 1960s that was hosted by Monty Hall. Consequently, the problem became known as the *Monty Hall problem*. It is one of those types of problems that are easy to state but that trip up many people, including, as we shall see, many professors of mathematics. Here's the problem:

Suppose you are a contestant on a TV show. You are given the following information: There are three sealed boxes, say, Box 1, Box 2, and Box 3. Two of the boxes are empty. You are informed that the third box contains one million dollars. You are not informed which box contains the loot. All you know is that one box certainly contains one million dollars. The host of the show informs you that *he* knows which box contains the million dollars.

You are asked to choose one box and one box only. If that box contains the cash, then you get to keep the million dollars. If you choose an empty box, then I am afraid you do not win any cash at all.

Now it is time for you to choose a box.

Say you choose Box 1.

Having made your choice, the TV host then opens one box (say, Box 3, but definitely NOT the box you chose) and shows that it is empty. The host then offers you a choice: You can stick with your initial choice (Box 1) or change your choice to Box 2. What should you do?

The commonsense approach would probably be to stick to the original choice. After all, you have made your choice. It is a guessing game. Why change now? Most people probably would not change. They would argue that it is a fifty-fifty chance (between Box 1 and Box 2) that either box contains the cash. They would probably maintain that swapping does not make any difference. So probably most would stick with their original choice.

When this puzzle is put to the ordinary man or woman on the street, most of them appear to answer that it does not matter whether or not you switch your choice. They believe that the contestant is faced with a fifty-fifty situation, and therefore there is an equal probability of the cash being in either box.

That is probably what most people would think. But here swapping makes all the difference. By changing boxes in this example, one *doubles* the chances of gaining the million dollars. Before reading any further, can you figure out why changing boxes is the sensible thing to do?

SOLUTION

The key to the problem with the three sealed boxes lies in the fact that the host knows which box contains the million dollars *before* he opens the box.

Ok. Let the game begin. You choose one box, say, Box 1. The chance that Box 1 contains the one million dollars is clearly 1 chance in 3, or 1/3. The host then opens one of the other boxes, say, Box 3, and shows that it is empty. (It is important to remember that the host knew Box 3 was empty *before* he opened it.)

There are now only two sealed boxes, 1 and 2. We have established that the chance that the million dollars is in Box 1 is 1/3. Therefore, the chance that the million dollars is in Box 2 (the unopened box) must be 2/3. In other words, it must be twice as likely that the million dollars is in Box 2 as in Box 1. You should therefore change your original choice. This reasoning applies no matter which box was originally chosen, or which box was opened by the host.

When this solution is presented to many people, they still find it incredible that changing from the original choice increases the probability of selecting the box with the cash in it.

Here is a more dramatic way of seeing why switching is the best procedure to adopt.

Suppose that instead of just three boxes, there are one million boxes on the stage. Each of the boxes is numbered from one to a million. You are informed that only one box contains the million dollars. You are told that all the other boxes (all 999,999 of them) are empty. You are also informed that the host of the show knows precisely which box contains the one million dollars. You are asked to choose one of the million boxes.

You make your choice. Say you pick Box 1. The chance that you picked the box containing the one million dollars is tiny. In fact, it is precisely one chance in a million. Therefore, the chances that you have chosen the box which does NOT contain the cash are 999,999 in a million. In other words, it is most likely that you have chosen a box that does not contain the cash. In fact, the probability that you have chosen an empty box is 0.999999. In contrast, the probability that you have chosen the one box containing the cash is 0.000001.

The game continues. The host now proceeds to open all the other boxes, except one (say, Box 500,000) and shows that all the boxes he has opened are empty. He then offers you a choice: stick to your original choice (Box 1) or choose Box 500,000.

At this point there are only two unopened boxes on the stage. The box that you chose and the box numbered 500,000. The chances that you chose the box containing the cash is still one chance in one million. Therefore, the chance that Box 500,000 contains the cash must be 999,999 chances in one million. It is almost certain that the cash is in Box 500,000. Therefore, you should definitely change from your original decision.

To solve the puzzle containing just three boxes, your reasoning can proceed as follows: It makes sense to change your original choice if there are one million boxes; or if there are 999,999 boxes; or if there are 999,998 boxes, and so on, down to just three remaining boxes. Consequently, if there are three boxes or more, it makes sense to change your original choice.

Here is yet another way of seeing the solution.

If there are three boxes to choose from, the money can be in only one of those three boxes: Box 1, Box 2, or Box 3.

Let us suppose the money is in Box 1.

If the contestant chooses Box 1 and switches, he loses.

If the contestant chooses Box 2 and switches, he wins.

If the contestant chooses Box 3 and switches, he wins.

By switching, the contestant wins in two out of three cases. The same reasoning applies if the money is in Box 2 or Box 3.

Graphically one can see this by studying the following table, which illustrates the results if the contestant chooses one box, say, Box 1.

Hidden behind Box 1	Hidden behind Box 2	Hidden behind Box 3	Result of Not Switching	Result of Switching
One Million Dollars	Nothing	Nothing	Win	Lose
Nothing	One Million Dollars	Nothing	Lose	Win
Nothing	Nothing	One Million Dollars	Lose	Win

One can see from this that the contestant who switches is likely to win two times out of every three. The contestant who chooses not to switch will lose two times out of three.

Here's yet *another* way of looking at the problem. The contestant is told at the start of the game that after he makes his initial choice, the host is going to open one box that will contain nothing. The contestant will then be given a choice: either stick with his original choice or switch to the second unopened box.

At this stage of the game, the contestant could reason as follows:

I am going to pick one box now. After I have made my choice the host is going to do something that is equivalent to giving me a choice of two boxes, one of which (he will demonstrate) is empty. When that is done, I still will not know for certain which box contains the money. However, it is better to accept the host's invitation to switch boxes, because my chances of getting the money will obviously increase if I am able to choose two boxes instead of one. Therefore, no matter which box I initially choose, I should switch.

One more example to bring home that switching is the right procedure: Suppose the game show runs along the following lines. At the start of the game, the contestant is informed that she will be asked in a moment to choose one card in a deck of fifty-two facedown cards. The contestant is further told that if she picks the queen of hearts, she will win one million dollars. If the contestant chooses any other card, she loses. The contestant is also told that the host of the show knows the *exact* position of the queen of hearts in the deck.

The game begins. All of the cards are shown facedown. The contestant chooses the twenty-first card from the top. She takes the card, still holding it facedown and places it facedown on a nearby table.

The host of the show now takes the remaining fifty-one cards and turns them over one by one, revealing that none of them is the queen of hearts. The host turns over every card, except for the second last card of the deck. He leaves this one card facedown.

The contestant is now asked if she wishes to switch from the card she initially chose to the one card that the host left unturned. What should the contestant do?

I think the answer must be obvious by now. She should SWITCH, of course!

The Monty Hall problem achieved national fame in the United States in September 1990. At the time, Marilyn vos Savant ran a column, "Dear Marilyn," in *Parade*. In the weekly column, she answered readers' questions concerning puzzles.

Let me point out that Marilyn vos Savant comes with an excellent pedigree. She was at one time the world record holder in *The Guinness Book of World Records* for having the highest recorded IQ in the world. Amazingly, Marilyn's IQ was set at 228.

In September 1990, a reader of the magazine named Craig F. Whitaker, from Columbia, Maryland, wrote to Marilyn to ask if it is to the advantage of the contestant to switch when participating in the puzzle game known as the Monty Hall problem. Marilyn correctly answered, saying it most certainly is to the player's advantage to switch. In fact, Marilyn maintained that the contestant who switched doubled his chances of winning the million dollars.

Soon after Marilyn published the correct solution, she received thousands of letters from members of the general public and from prominent professional mathematicians in various universities in the United States. The vast majority of these letters from the public and from the mathematicians argued that Marilyn's answer was wrong. Unfortunately, many of the letters from the mathematicians were very critical of Marilyn and verged on being abusive.

One professional mathematician, who obviously was very annoyed with Marilyn, went so far as to suggest that Marilyn refer to a standard textbook on probability before attempting to solve a similar problem again.

Another professional mathematician said that Marilyn was "utterly wrong," and publicly hoped that the controversy caused by Marilyn's answer would call attention to the deplorable state of affairs in mathematical education in the United States. He further suggested that if Marilyn would admit her error, she would contribute constructively toward the solution of a deplorable situation. He went on to ask how many irate mathematicians were needed for Marilyn to change her mind.

Yet another mathematician stated that he was sure that Marilyn would receive many letters from high school and college students concerning this problem and suggested that Marilyn keep their addresses in case she might need to contact them in the future for help with further problems.

Marilyn even received letters from the deputy director of the Center for Defense Information, and a research mathematical statistician from the National Institutes of Health. Both strongly argued that Marilyn's answer was wrong.

Marilyn has stated that about 92 percent of the letters from the general public maintained that her answer to the problem was incorrect. Of the letters Marilyn received from the universities, about 65 percent argued that Marilyn's solution was incorrect. Generally, of all the letters relating to the problem received by Marilyn, 90 percent stated that her answer was wrong.

Marilyn eventually asked schoolchildren, high school students, and college students throughout the United States to try a little experiment in their schools and colleges. Marilyn suggested that the children and adults play the game and to check their results.

They did just that.

Many of these children and their teachers, the high school students, and college students subsequently wrote to Marilyn stating that their experiments confirmed that the contestant who switched won on average about twice as often as those contestants who chose not to switch.

These results, as many of the children were able to confirm, are in accordance with probability theory.

There are many problems in probability theory that have unexpected and surprising solutions.

These problems are usually framed as a bet to an unsuspecting customer in a bar or at a fairground attraction. These types of bets are often referred to as *proposition bets* by con artists. The bet *appears* to be advantageous to the person being presented with the bet. That person is usually referred to as "the sucker" or "mark" by the con man. Hence these types of bets are often referred to as *sucker bets* as well.

One of the best sucker bets goes as follows: The con man presents you with three cards, say cards A, B, and C. One card, say, Card A, has the color blue on both sides. The second card, Card B, has the color red on both sides. The third card, Card C, has blue on one side and red on the other.

The con man asks you to shuffle the three cards and then asks you to place the three cards inside a box. You both leave the room, and a third person picks one of the three cards from the box and places it on the table. Both you and the con man return to the room. You both see that the top color of the card on the table is red.

Pointing at the card on the table, the con man says to you: "I am willing to bet you twenty dollars at even money that the other side of this card is red."

Should you take the bet?

You may reason as follows: This card on the table cannot be the card that is colored blue on both sides. So we can eliminate that card. Therefore, the card on the table is either the card colored blue on one side and red on the other side, or it is the card colored red on both sides. It is either one card or the other. Therefore, the odds must be even that the other side of the card is red and that the other side of the card is blue. So it is an even-money bet.

Of course, this is exactly how the con man wants you to reason. The truth is that the other side of the card in two of three cases is red.

Think of it this way. The card on the table colored red can only be one of two possible cards. Either it is the card colored red on either side or the card colored red on one side and blue on the other.

In our minds, let us imagine the number 1 printed in invisible ink on one side of the card colored red and number 2 printed on the other side, also colored red. Let us also imagine the number 3 printed in invisible ink on the red side of the other card and let us imagine the number 4 also printed in invisible ink on the blue side.

Consider the card on the table that is colored red on top. That card can be side 1 or 2 of one card, or side 3 of the second card. In two of these cases, the other side of the card is red. In only one case is the other side of the card blue. Therefore, there are 2 chances in 3 that the other side of the card is red. When the con man offers you the bet with odds of even money, he is shortchanging you. He will win the bet 2 times out of every 3.

Here is an old but still quite a good sucker bet. In a bar, a stranger may wander up to you and ask you to inspect and then shuffle a normal deck of fifty-two cards. When you have shuffled the deck, he will ask you to look through the deck and remove the first picture card you see.

Let's assume this card is a king. He will then ask you to look through the deck a second time and remove the first picture card you see that is not a king. Suppose this second card is a queen. The con man will then ask you to look through the deck a third time and remove the first picture card that is not a king or queen. In other words, he will ask you to remove the first jack you see.

Before you actually do that, however, he bets you twenty dollars at even money that two of these three cards, the king, queen, or jack, will be of a similar suit. Should you accept the bet?

If you are a person who likes to gamble, you will probably reason as follows: There are four kings, four queens, and four jacks in a normal deck of cards. That is a total of twelve picture cards in the deck. This stranger (the con artist) has asked me to select the first king, queen, and jack that I see in the deck. It is highly unlikely that two of those three cards will be of the same suit. Therefore, the bet is my favor and I should accept the wager. It will be a handy way of picking up twenty bucks.

Of course, this is how the con man wants you to reason!

The odds that two of the cards are of a similar suit are 5 to 3. This is how these odds are calculated:

There are 4 kings, 4 queens, and 4 jacks in a deck of cards. That means that there are $4 \cdot 4 \cdot 4$ or 64 different ways the 12 picture cards can be arranged.

The order of the three cards drawn does not alter the odds in any way. Therefore, let us assume the first picture card drawn is a king. The king can be any one of 4 suits. If a queen is drawn next, it can be a different suit from the king in 3 ways. When a jack is drawn, it can be a different suit from both the king and the queen in 2 ways.

Therefore, the total number of ways the first king, the first queen, and the first jack can be three different suits is $4 \cdot 3 \cdot 2$. This equals 24 ways. But there are 64 ways of arranging the kings, queens, and jacks. Therefore, there must be 40 ways that at least two of the three cards are of a similar suit. Thus the odds that two of the three cards are of a similar suit are 40 to 24, or 5 to 3. In other words, there are 5 chances in 8 that two of the three cards will be of a similar suit. The advantage is with the con man. Do not accept the bet.

I sent a copy of this chapter to Dr. Cong and asked for any comments or observations he may wish to make. He sent the following e-mail to me.

> Hi, Owen,
>
> I read your piece on the Monty Hall problem and the other tricky probability puzzles. I found it interesting. The Monty Hall problem is a puzzle that still stumps people when they are presented with it. You gave a comprehensive review of the puzzle, so I will sum up the situation the contestant faces in as few words as possible: At the beginning of the game show, the contestant has a 1/3 chance of picking an empty box. Therefore, by switching she literally switches the odds of winning to 2/3.

Here is another sucker bet with cards that your readers may like.

The con man takes two kings, two queens, and two jacks from a deck and shuffles the six cards. Then he places each of the cards facedown in a row on a table. He points out to you that of the six facedown cards on the table, two of those are queens. He now offers you the following bet. He asks you to choose any two cards from the six facedown cards on the table. He places twenty bucks on the table and offers odds of 10 to 1 that you will not succeed in choosing two queens.

You may well be tempted with this bet. Perhaps you would reason as follows: The odds are against me because two thirds of the cards on the table are not queens. Therefore, the chances are one in three of selecting the pair of queens. This equates to two in three of not picking the pair of queens. But this guy (the con artist) is giving odds of 10 to 1 that I will not choose the two queens. I think I will take this bet. If I am lucky and I win, I will win two hundred dollars. If I lose, I will bet a second time, and if I lose that, I will bet a third time. I am likely to win one bet out of the three. If that happens, I will lose forty dollars but I will win two hundred dollars. That is a nice profit of one hundred and sixty dollars. The bet surely is in my favor. So I should accept the bet.

If you do, then you have played right in to the hands of the con man!

Let's analyze the bet a little closer. There are six cards on the table. Two cards are queens. What is the probability that if you choose two cards that those two cards will be queens? The first card you choose can be any one of six cards. The second card you pick can be any one of five cards. That is $6 \cdot 5$ or 30 ways of picking two cards from six cards.

But that calculation does not differentiate between the *order* of the two chosen cards. In other words, if the two cards you picked are card A and card B, the selection A and B is counted and so is the selection B and A. These count as two selections. This procedure counts the number of *permutations*. But if we choose to ignore the order, we treat A and B as the same selection as B and A. Therefore, these two selections count as one selection. This procedure counts the number of *combinations*. To get the number of combinations, we multiply 6 by 5, getting 30, and then divide 30 by 2, obtaining 15. Therefore, there are 15 ways of selecting two cards from six, ignoring the order of the two selected cards. In other words, there are 15 *combinations* of six objects taken two at a time.

Thus the chance that you will choose two queens is 1 chance in 15. The correct odds of picking a pair of queens from the six cards are 14 to 1.

In other words when the con man offers you odds of 10 to 1, he is shortchanging you. You will win this bet only once in every fifteen tries over the long term. Therefore, the con man can expect to win this bet 14 times in every 15 bets. Those are handy odds for the con man!

So it is highly likely you will be parted from a number of your greenbacks if you accept this bet.

Here is one final sucker bet:

The con man will take a normal deck of cards and ask you to shuffle them. He will then ask you to cut the deck into three heaps. The con man then bets you twenty dollars at even money that at least one card on top of the three heaps is a king, a queen, or a jack. (The kings, queens, and jacks are known as picture cards, or court cards.) Should you accept the bet?

Absolutely not!

There are $\frac{52 \cdot 51 \cdot 50}{6}$, or 22,100 ways three different cards can be chosen from 52. Many of these selections (but not all) will include three cards that may consist of one or two or three picture cards.

If one removes the twelve picture cards from the deck, there will be 40 cards remaining. The number of ways of choosing three cards from 40 is $\frac{40 \cdot 39 \cdot 38}{6}$, or 9,880. All of these 9,880 selections will NOT include any kings, queens, or jacks. Let us call this group of 9,880 sets of three cards "Set A."

Since there are a total of 22,100 sets of three cards in a deck, and 9,880 of these sets do not contain a picture card, then (22,100 – 9,880) or 12,220 sets of three cards in a deck (of fifty-two) must contain at least one picture card. Call these 12,220 sets of three cards "Set B."

Let's now return to the problem of the deck being cut into three heaps. Consider the top card of each of the three heaps. Those three cards is a set of three cards that must belong to either "Set A" or "Set B."

If those three cards belong to "Set A," it is impossible that one or two or three of the cards is a picture card.

If those three cards belong to "Set B," then at least one of the three cards is a picture card.

So the probability that any one of the three cards turned up is NOT a picture card is 9,880/ 22,100, which equals 44.705 percent.

Therefore, the probability that at least one of the three cards turned up is a picture card is 12,220/22,100 or 55.294 percent.

Therefore, the con man has the advantage. The odds in this bet are approximately 55 to 45 in his favor. Do not accept this bet.

Kind regards,

Dr. Cong

CHAPTER 5

THE FIBONACCI SEQUENCE

The sequence we now call the *Fibonacci sequence* was named by the French mathematician Édouard Lucas (1842–1891), because an Italian mathematician named Leonardo Bonacci (ca. 1170–ca. 1250) studied the sequence. Bonacci, who was the son of a wealthy Italian merchant, wrote a book in 1202 titled *Liber Abaci* (Book of Calculation). It is in this book that Bonacci first mentions the sequence. It is believed that he had obtained the sequence from Indian merchants. Scholars today believe that the sequence was known in India in the sixth century.[1] Bonacci is much better known to the world of mathematicians today as "Fibonacci," which is believed to mean "son of Bonacci." It was by way of this book that Fibonacci introduced European mathematicians to the sequence now named after him. Later, Lucas discovered his own sequence—similar in many ways to Fibonacci's—which bears his name: the Lucas numbers.

The Fibonacci sequence is as follows: 1, 1, 2, 3, 5, 8, 13, 21, 34, 55, 89, 144, 233, 377, 610 . . . each successive number after the second term being the sum of the two previous numbers.[2] The *n*th number in the sequence is denoted as F_n. Thus $F_1 = 1$, $F_2 = 1$, $F_3 = 2$, $F_4 = 3$, $F_5 = 5$, $F_6 = 8$, and so on. The sequence was published in Fibonacci's *Liber Abaci* in 1202 as a solution to a trivial puzzle about the number of pairs of rabbits bred over the course of a year.[3] The sequence did not initially attract great interest. It was not until the beginning of the nineteenth century that other mathematicians began to study the sequence. Lucas was one of those mathematicians, and he named the sequence, 1, 1, 2, 3, 5, 8 . . . the *Fibonacci sequence*.

The sequence today is known by most mathematicians because of the many beautiful properties found within it. But the sequence also has a practical use. In recent years, it has been found to be extremely helpful to computer programmers in computer data storage and processing. (This is one of the strange consequences of mathematical discoveries. Often mathematicians will study some pattern in mathematics that has no apparent use in the physical world. They study the pattern mainly because they find the mathematical properties of the pattern interesting in themselves. Then, years later, sometimes centuries later, as in the case of the Fibonacci sequence, mathematicians find that the pattern that they have studied for so long has indeed a practical use that no one could have foreseen.)

Many students of mathematics are astonished when they first encounter some of the properties of the Fibonacci sequence. The following are just some of these amazing identities.

Suppose one wishes to find the sum of the squares up to the *n*th Fibonacci number. This sum equals the product of F_n and $F_{n+1.}$ Hence the sum of the squares of the first five Fibonacci numbers is F_5 times F_6, which works out as 40. Every *k*th number in the Fibonacci sequence is a multiple of F_k. Thus every third number is a multiple of 2, every fourth number is a multiple of 3, every fifth number is a multiple of 5, every sixth number is a multiple of 8, every seventh number is a multiple of 13, and so on.

Mathematicians discovered these properties—and many more—because they searched for properties in the sequence.

What made them look for properties in the sequence in the first place?

Well, if any sequence of numbers comes to the attention of a mathematician—for any reason—he will instinctively search for any surprising properties that may not be initially discernible. Given the fact that the Fibonacci sequence appears in various natural phenomena, this would only encourage the mathematician to look even deeper into the secrets of the sequence. There is a belief among many mathematicians that if nature "chooses" to use a particular pattern of numbers in its manifestations, then there must be something special about the sequence. This attitude and philosophy has proved fruitful down through the centuries, as we know only so well from studying the Fibonacci sequence.

This philosophy that many mathematicians have brings us close to the heart of the question of what mathematics *is*. Is it a game that mathematicians play by using symbols and various rules that we *invent*? Or is mathematics some kind of universal set of rules that exist "out there," in an abstract way, whose rules we cannot alter; indeed, whose rules nature appears to follow; and that by following these rules we *discover* the results that this mathematical reality imposes upon us?

This question is still hotly debated among mathematicians and philosophers. So we will not debate it any further here.

The sum of the squares of two consecutive Fibonacci numbers is another Fibonacci number. For example, $5^2 + 8^2 = 89$. This is the eleventh Fibonacci number. Or $8^2 + 13^2 = 233$. This is the thirteenth Fibonacci number. Or $13^2 + 21^2 = 610$. This is the fifteenth Fibonacci number. Or $21^2 + 34^2 = 1{,}597$. This is the seventeenth Fibonacci number. And so on. In general, $F_n^2 + F_{n+1}^2 = F_{2n+1}$.

Consider any three consecutive Fibonacci numbers. Let's take 2, 3, and 5. The sum of $3^3 + 5^3 - 2^3 = 144$ $(27 + 125 - 8 = 144)$. This equals the twelfth Fibonacci number. The result of this procedure (144 in this case) will always be a Fibonacci number. For example, consider the seventh, eighth, and ninth Fibonacci numbers. These are 13, 21, and 34. We find that $21^3 + 34^3 - 13^3 = 46{,}368$ $(9{,}261 + 39{,}304 - 2197) = 46{,}368$. It so happens that 46,368 is the twenty-first Fibonacci number. This property will always hold. In general, $F_n^3 + F_{n+1}^3 - F_{n-1}^3 = F_{3n}$.

There is no Fibonacci number greater than 8 that is one less or one greater than a prime number. The only square numbers in the Fibonacci sequence are 1 and 144. Curiously, 144 is the twelfth Fibonacci number, and 144 is also the square of 12. The fourth Fibonacci number (usually written F_4) equals 3, which is prime, but F_4 has a composite subscript. This is a unique case. Every other prime in the Fibonacci sequence has a prime subscript. The converse is not true, however. If a number has a prime subscript, it does not necessarily mean that that Fibonacci number is prime also. The nineteenth Fibonacci number (F_{19}) has a prime subscript, but that Fibonacci number is 4,181, which equals 37 times 113. The next Fibonacci number that is composite but has a prime subscript is F_{31}. It equals 1,346,269, which equals 557 times 2,417.

If, and only if, an integer, *m*, divides evenly into an integer, *n*, then F_m divides F_n. For example, 6 divides evenly into 24. Therefore, F_6, which equals 8, divides evenly into F_{24}, which equals 46,368 (8 divided into 46,368 equals 5,796). Or, for example, 7 divides into 21. Therefore, F_7, which equals 13, divides evenly into F_{21}, which equals 10,946. And we find that 13 does divide evenly into 10,946. It goes in 842 times.

The greatest common factor of two Fibonacci numbers is always another Fibonacci number. For example, the greatest common factor of 13 and 55 is 1; the greatest common factor of 8 and 34 is 2; the greatest common factor of 144 and 2,584 is 8; the greatest common factor of 610 and 6,765 is 5; and so on. All of these common factors are Fibonacci numbers.

The only powers of two in the Fibonacci sequence are 1, 2, and 8. The only cubes in the Fibonacci sequence are 1 and 8.

It is not known if there are an infinite number of Fibonacci numbers that are prime, although it has been conjectured that there are.[4] The position with Lucas numbers that are prime is similar. The Lucas numbers are: 1, 3, 4, 7, 11, 18, 29, 47, 76 . . . . The sequence begins with 1 and 3, and thereafter each term is the sum of the preceding two numbers. It is widely believed by mathematicians that there are an infinite amount of Lucas primes, but no one has yet proved this conjecture.[5]

The Fibonacci sequence turns up frequently in various ways in nature. Many of these appearances have been well documented in books and magazine articles over the years. Therefore, I will mention just two simple examples.

The seeds on the head of all sunflowers are arranged in two spirals, one going clockwise, the other counterclockwise. In the vast majority of cases, the number of seeds in each spiral is two consecutive Fibonacci numbers. For example, quite often the number of seeds in one spiral will be 34, and in the opposite spiral 55. Less frequently, one will find 55 seeds in one spiral and 89 seeds in the opposite spiral. To see if this tendency continues, giant sunflowers have been grown. In these rarer cases, 89 seeds are found to be going one way and 144 seeds are found to be going in the other direction.

Why does nature always choose to have the number of seeds in spirals going clockwise and counterclockwise to be two consecutive Fibonacci numbers? Apparently botanists find that this is the optimal way to fill the space on the head of a sunflower. Sunflowers have evolved in such a way that they optimize this space. This is yet another example of mathematicians finding a pattern that has interesting mathematical properties, and then discovering at a much later stage that this pattern appears in nature also.

The second example of the Fibonacci sequence appearing in nature concerns the ancestry of an idealized male bee. It may come as a surprise to some readers that a male bee (commonly known as a drone) comes from an unfertilized egg. If the egg is fertilized, the offspring is female. (See figure 5.1.)

Let us examine a hypothetical situation where we trace the ancestry of an idealized male bee. In this scenario, we assume each male bee reproduces one bee (a female) and each female bee reproduces one female and one male bee.

At the bottom of the illustration in figure 5.1, we have one male bee (MB). In this generation, there is only one bee that we are considering.

That male bee produces one offspring: a female bee. So in the second generation we also have one bee. The female bee in the second generation produces two offspring: a female bee and a male bee. So in the third generation there are two bees. The female bee in the third generation produces two offspring: a female bee and a male bee. The male bee in the third generation produces a female bee. So in this fourth generation there are three bees. We then find that these three bees produce five offspring in the next generation, and so on.

The illustration in figure 5.1 shows us that a male bee has only one parent: its mother. It has two grandparents (its mother's parents), three great-grandparents (recall its mother's father had no father), five great-great-grandparents, and so on. As one goes back through the ancestry of a male bee, the Fibonacci sequence emerges!

FB MB	FB	FB MB	→	5 Bees in This Generation
↑	↑	↑		
FB	MB	FB	→	3 Bees in This Generation
↑		↑		
FB		MB	→	2 Bees in This Generation
	↑			
	FB		→	1 Bee in This Generation
	MB		→	1 Bee in This Generation

Figure 5.1

The Fibonacci sequence also turns up in real everyday situations that you may encounter. Suppose you are in an antique store and wish to purchase beautiful miniature replicas of, say, the Statue of Liberty. Let's assume that each replica costs ten dollars. Suppose your wallet contains a substantial number of ten-dollar and twenty-dollar bills only. Suppose you want to spend ten dollars to purchase one replica. There is only one way of doing it. Hand over the ten-dollar bill for the item you wish to purchase. Suppose you want to spend twenty dollars. There are two ways of doing that: spend a ten-dollar bill and then spend another ten-dollar bill, or else simply spend one twenty-dollar bill.

Suppose you decide to spend thirty dollars. You will be able to do this in three ways: You can first spend three ten-dollar bills; or one ten-dollar bill plus a twenty-dollar bill; or a twenty-dollar bill plus a ten-dollar bill. If you wish to spend forty dollars, you will find that there are five ways of doing it: four ten-dollar bills; two ten-dollar bills plus a twenty-dollar bill; one ten-dollar bill plus a twenty-dollar bill plus a ten-dollar bill; a twenty-dollar bill plus two ten-dollar bills; or simply two twenty-dollar bills.

Similarly, you will find that there are eight ways of spending fifty dollars; thirteen ways of spending sixty dollars, and so on. You will notice that the number of ways that ten dollars, twenty dollars, thirty dollars, forty dollars, and so on, can be spent is 1, 2, 3, 5, 8, 13 . . . . This is the Fibonacci sequence beginning with the second term of the sequence.

The *n*th Fibonacci number can be found by the following formula:

$$\frac{((1+\sqrt{5})^n)}{2^n} - \frac{((1-\sqrt{5})^n)}{2^n} / (\sqrt{(5)})$$

The formula can be awkward to use for large values of n. But if we make n equal to 2, we can see the formula in action: The formula then becomes:

$$\frac{((1+\sqrt{5})^2)}{2^2} - \frac{((1-\sqrt{5})^2)}{2^2} / \sqrt{5}$$

$$\frac{((3.2360679...)^2)}{2^2} - \frac{((1.2360679...)^2)}{2^2} / (2.2360679...)$$

$$\frac{2.61803398... - 0.38196601}{2.23606797} = 1$$

The formula tells us that the second Fibonacci number is 1, which is correct. You should bear in mind that this formula was used by mathematicians prior to the days of scientific calculators. At that time it was tedious to calculate large Fibonacci numbers, when a mathematician needed to know a particular Fibonacci number without having to calculate all of the previous Fibonacci numbers. That is where the above formula became useful. With the widespread availability of scientific calculators, present-day number fans do not require this formula to calculate a particular Fibonacci number. (To find the *n*th Fibonacci number, raise the number 1.61803398 . . . to the power of n, divide the result by the square root of 5, and round to the nearest integer; the answer is the nth Fibonacci number.) As is the case of a number of other mathematical procedures (such as the calculation of logarithms), the formula has become an obsolete curiosity. However, modern-day recreational mathematicians often like to learn of these formulae, if for no other reason than that it helps them to appreciate the marvels of today's scientific calculators.

Many students of recreational mathematics are pleasantly surprised to learn that the sum of the consecutive Fibonacci numbers from F_1 to F_n equals $F_{n+2} - 1$. For instance, the sum of the first twelve Fibonacci numbers equals $F_{14} - 1$, or 376. In other words, $1 + 1 + 2 + 3 + 5 + 8 + 13 + 21 + 34 + 55 + 89 + 144 = 376$. The fact that these unexpected connections exist gives the student studying the Fibonacci sequence reasons to believe that there is an underlying unity and order in mathematics.

The Fibonacci sequence is related to the Pythagorean theorem in a surprising way. (Mathematicians are always delighted to find connections between two different mathematical entities that were originally assumed not to be related. It reinforces their belief in an underlying unity in mathematics.) Take any four consecutive Fibonacci numbers: say, 2, 3, 5, and 8. Multiply the two outside numbers together (2 and 8) to obtain 16. Then double the product of the two inside numbers, 2 times $(3 \cdot 5)$ to obtain 30. Then 16 and 30 are two legs of a Pythagorean right triangle: We find that $16^2 + 30^2 = 34^2$.

As a bonus, the *hypotenuse* of such a triangle will always be a Fibonacci number. In this case, the hypotenuse is 34, which is the ninth Fibonacci number.

The *area* of such a triangle will equal the product of the relevant four consecutive Fibonacci numbers. In our example, the area of the 16, 30, 34 right triangle is 240, which equals the product of 2, 3, 5, and 8. The *perimeter* of such a triangle equals twice the product of the last two relevant consecutive Fibonacci numbers. In the case cited here, the perimeter of the 16, 30, 34 triangle is 80, which is just twice the product of 5 and 8.

There is a simple method to discover if a positive integer, x, is a Fibonacci number. If one or both of $5x^2 + 4$ or $5x^2 - 4$ is a square integer (call it y), then x is a Fibonacci number. As a bonus, y will be the square of the corresponding Lucas number! (The reader may recall reading earlier in this chapter that the Lucas sequence is 1, 3, 4, 7, 11, 18, 29, 47, 76. . . . We will briefly discuss Lucas numbers in the next chapter.) Suppose you want to check if 987 is a Fibonacci number? To check, calculate five times the square of 987, and add 4 to the answer. This equals 4,870,849, which is the square of 2,207. Therefore, 987 is a Fibonacci number.

Actually, 987 is the sixteenth Fibonacci number, and 2,207 is the sixteenth Lucas number. The fact that a check such as this exists to discover if a positive integer is a Fibonacci number is surprising in itself. The fact that the procedure involves the squares of Lucas numbers is totally unexpected.

The Fibonacci numbers sometimes turn up in the solutions of combination problems that may arise from day to day. Here is one such puzzle: A feeble man has to climb to the top of a staircase consisting of n steps. He is only able to take one or two steps at a time. How many distinct ways are there of climbing the stairs in this manner? The solution, which is always a Fibonacci number regardless of the number of steps to the top of the stairs, is given at the end of this chapter.

I sent Dr. Cong a copy of this chapter to ponder in his retirement on the West Coast of the United States, but he has lost none of his marvelous insight into numbers. I asked him if he would furnish me with some interesting facts about the Fibonacci numbers, and his response is as sharp as ever.

Hi, Owen,

I found your discussion of the Fibonacci numbers interesting. Here is a simple identity that applies to the Fibonacci sequence. Consider any four consecutive Fibonacci numbers. Take the first four: 1, 1, 2, 3. Square the first and fourth Fibonacci and sum the results. We obtain: $1^2 + 3^2 = 10 = 2 \cdot 5$. The answer, in this case, 10, is twice a Fibonacci number. This is always the case. Consider the following four consecutive Fibonacci numbers: 1, 2, 3, 5. Sum the squares of the first and fourth Fibonacci numbers, obtaining 26, which equals twice 13. And so on.

I like the examples of situations in which the Fibonacci sequence unexpectedly turns up. Your readers might like to learn of another unexpected place where the Fibonacci sequence makes an appearance. Suppose you write down a string of n digits containing only 0s and 1s, so that no 0s are adjacent to each other. If n equals zero, it is impossible to write down any digits. Therefore, there is only one way to write down a string of digits containing zero digits.

Let's say n equals one. That means we are dealing with just one digit. So we write down one single 0 or one single 1. They are the only 2 possibilities. If n equals two, there are 3 possibilities. You many write down (0, 1,), (1, 0) and (1, 1). If n equals three, there are 5 possibilities: (0,1,0), (1,1,0), (1, 0, 1), (0,1,1), and (1,1,1,).

If the reader cares to check, she will find that when the string of digits equals four, there are 8 possibilities; and when the string contains five digits, there are 13 possibilities; and so on. Your readers will notice that the Fibonacci sequence emerges in the number of ways of writing n digits so that no two 0s are adjacent to each other, as n gets larger and larger.

The Fibonacci sequence is a marvelous sequence. In my opinion, every child on Earth should be taught the Fibonacci sequence. It is one wonderful way of opening a child's mind to the wonders of mathematics and to the many wonders of nature.

Children and teenagers are naturally curious. I think teachers should keep this in mind when teaching mathematics. Teachers should encourage children to study number sequences in order to find interesting patterns. Let them know that many mathematicians do this every day. Don't let the fact that there may not be a practical use for the sequence that we know of put the kids off learning about it. Sometimes a practical use is found that involves the pattern that is being discussed. Sometimes there isn't. At least not yet! But who is to say that there will not be a practical use for the pattern found sometime in the future? The history of mathematics is full of such discoveries.

If I were teaching children or teenagers in school today, I would try to introduce a little fun into their lessons. I would explain to them that the pattern 1, 1, 2, 3, 5, 8, 13, 21, 34, 55, 89, 144, 233 . . . is one of the most fascinating number patterns that human beings know of. I would encourage the children to realize that the Fibonacci sequence is worth knowing.

In order to arouse their interest, I would point out the following:

I would mention that because a mile equals 1.609347 + kilometers, the Fibonacci sequence may be used to give a good conversion from miles to kilometers, or vice versa. For example, 3 miles is reasonably close to 5 kilometers; 5 miles is reasonably close to 8 kilometers; 8 miles is reasonably close to 13 kilometers, and so on.

I would ask the children to look at their two hands. I would point out to them that they have 1 thumb on their right hand and 1 thumb on their left hand. Those are the first two numbers of the Fibonacci sequence. Then I would point out that they have 2 bones in each of their thumbs; 3 bones in each of their fingers; 5 digits on each of their hands and a total of 8 fingers. Those numbers, 1, 1, 2, 3, 5, 8, are the first six Fibonacci numbers.

I would let the kids know that the Fibonacci numbers are related to their healthy snacks. For example, inside the skin of a banana we will find 3 sections; and if we cut an apple in two, we find that it has 5 sections. Both 3 and 5 are Fibonacci numbers.

I would ask the children to write down the first ten integers. Then I would ask them to partition the integers, beginning with 1, to include only odd integers, including the number itself, if it is odd. These are the results for the first six integers:

Number to Be Partitioned	Partitions	Number of Partitions
1	(1)	1
2	(1,1)	1
3	(3); (1, 1, 1)	2
4	(1) (3); (3) (1); (1,1,1,1);	3
5	(5); (1, 1, 3); (1, 3, 1); (3, 1, 3) (1, 1, 1, 1, 1)	5
6	(5, 1); (1, 5); (3, 1, 1, 1); (1, 3, 1, 1); (1, 1, 3, 1); (1, 1, 1, 3); (1, 1, 1, 1, 1, 1); (3, 3)	8
	. . .	

Note that the number of partitions for the integers, 1, 2, 3, 4, 5, 6 . . . is 1, 1, 2, 3, 5, 8 . . . . These are the Fibonacci numbers.

Finally, I would tell the children that we are going to learn a little code. (Children usually like to learn about simple codes.) In this code, we give every letter of the alphabet a number. We will let the letter A equal 1; we will let the letter B equal 2; we will let the letter C equal 3; and so on, so that the final letter of the alphabet, Z, will equal 26.

I would then write the following two words: FIBONACCI SEQUENCE.

I would ask the children to substitute the correct number for each letter in that sentence. I would then ask them to add up those numbers. They should get 151.

I would then point out that $151 = 1 \cdot 12 + 35 + 8 \cdot 13$.

All the numbers in that equation are Fibonacci numbers.

I think that this approach to teaching young children in particular may help to open the children's eyes to the wonders of the Fibonacci sequence.

At least that would be my hope.

Here is one final curiosity before I sign off:

13 is the 7th Fibonacci number.
21 is the 8th Fibonacci number.
34 is the 9th Fibonacci number.

Note that 21 times (13 + 34) = 987, which is the 16th Fibonacci number.
All the best for now.
Dr. Cong

SOLUTIONS

(1) A feeble man has to climb a staircase to the top. The stairs consist of n steps. He is only able to take one or two steps at a time. How many distinct ways are there of climbing the stairs?

The solution can be achieved by reasoning as follows: Since he can get to the top of the stairs by taking one or two steps at a time, he can only reach the nth step from either the (n–1)th step or from the (n–2)th step. There is no other way he can climb to the top. Therefore, the total number of ways he can climb to the nth step is the sum of the number of ways he can climb to the (n–1)th step or the (n–2)th steps.

We can tabulate this result as follows:

Number of Steps in Stairs n	Total Number of Ways of Climbing n Steps, Taking One Step at a Time or Two Steps at a Time
0	1
1	1
2	2
3	3
4	5
5	8

. . .

One can see from the table that the solution to this puzzle involves the Fibonacci sequence. We see that if the number of steps in the stairs equals n, the total number of ways of climbing to the nth step equals $F_n + F_{n-1}$.

CHAPTER 6
THE LUCAS SEQUENCE

The Lucas sequence is named in honor of Édouard Lucas (1842–1891), a French mathematician who had studied the sequence now known as the Fibonacci sequence. It was he who gave the Fibonacci sequence its name.

The Lucas sequence is as follows: 1, 3, 4, 7, 11, 18, 29, 47, 76, 123, 199, 322, 521, 843, 1,364, 2,207, 3,571, 5,778, 9,349, 15,127, 24,476; the third term and all successive terms are the sum of the previous two terms. The *n*th term in the Lucas sequence is denoted as follows: $L_1 = 1, L_2 = 3, L_3 = 4, L_4 = 7, L_5 = 11, L_6 = 18, L_7 = 29$, and so on.

We do not know why Lucas began studying this sequence. Most likely he was curious—as most mathematicians are—about sequences. He had studied the Fibonacci sequence, where the first four terms are 1, 1, 2, and 3. He found that that sequence contained very surprising number properties. Consequently, he probably decided to see what properties may lay in a sequence if the first three terms of the sequence are 1, 3, and 4.

I have no doubt that Lucas believed—before he explored the sequence that was subsequently named in his honor—that curious properties already existed in some sense in the sequence. In other words, I have no doubt that Lucas viewed mathematical reality as existing outside human minds, and that any properties he should come across would be *discovered* by him, and certainly not invented by him.

There is nothing unusual in this approach. Pure mathematicians do not seek a reason to study any aspect of pure mathematics. They study mathematics for its own sake. They study mathematics because they believe it is a beautiful subject. The pure mathematician is not concerned whether her discoveries will be useful or practical in the everyday world. These thoughts are probably the farthest things from her mind.

No one apparently asks if basketball, baseball, or soccer is useful. Where did this idea that mathematics has to be useful come from? Is chess useful in the real world? I do not think so, but I do know that it gives tremendous pleasure to many folks around the world. Is baseball useful in the real world? Probably not, but there are millions who enjoy watching the game. The same goes for numerous other sports and hobbies.

When children are taught mathematics in school, it is usually taught with a view to giving the child a chance to become independent in life. The child will be taught in a progressive manner how to make change, measure lengths and areas, calculate her salary or wages that are due to her, and, hopefully, even calculate the amount of taxes she is legally obliged to pay the state.

When an individual is taught higher-level mathematics in a college it is usually because he wants to become an engineer, a scientist, a medical doctor, an accountant, an actuary, or any one of a dozen other professions.

That is all fine and well. However, there is a group of mathematicians, and indeed many ordinary folk, like you and me, who actually *enjoy* mathematics. I do not think anyone should have to explain why he enjoys mathematics, any more than anyone has to explain why he enjoys baseball.

Those who enjoy mathematics like the fact that mathematics gives consistent results. If today there twenty-five prime numbers less than one hundred, we know that next year there will still be only twenty-five primes less than one hundred. In fact, one thousand years from now there will still be only twenty-five primes less than one hundred.

If one thinks about it, mathematical truths, even simple ones, such as 1 plus 1 equals 2, or that 7 is a prime number, are more "real" and more "permanent" than the trees or buildings we see around us every day. Long after all those trees have died and long after all those buildings will have decayed and disappeared, 1 plus 1 will still equal 2 and 7 will still be a prime number. When you and I and the rest of the world's population are no more, mathematical truths will still mysteriously exist.

These pure mathematicians like the fact that even if a renowned professor of mathematics says that he has discovered a new mathematical theorem, that "theorem" will be granted only the status of a conjecture by the mathematical community. The credentials of the discoverer are irrelevant in deciding whether the discovery is actually true. The conjecture will be thoroughly checked for any errors that it may contain. Only when a relatively large number of suitably qualified mathematicians have gone through the conjecture, step-by-step, and are satisfied that the conclusion of the conjecture can be proved by a number of logical steps, is the conjecture considered proved and then given the status of a "theorem." (Recall, a mathematical theorem is true; a conjecture is just an educated guess.)

Similar to the Fibonacci sequence, the Lucas sequence has numerous interesting properties. Many of these have appeared in the literature over the years.

Before discussing some of these properties, it might be worthwhile for you to refer to the first twelve Fibonacci and Lucas numbers. (See figure 6.1)

Terms	Fibonacci Sequence	Lucas Sequence
1	1	1
2	1	3
3	2	4
4	3	7
5	5	11
6	8	18
7	13	29
8	21	47
9	34	76
10	55	123
11	89	199
12	144	322

Figure 6.1. The first twelve Fibonacci and Lucas numbers.

There are numerous connections between the Lucas numbers and the Fibonacci numbers. Here are just two of these.

Consider any number in the Lucas sequence. Say, the fourth Lucas number, which is 7. Multiply that by the fourth Fibonacci number, which is 3. The answer is 21, which appears in the Fibonacci sequence. Consider another Lucas number. Say, the fifth Lucas number, which is 11. Multiply that by the fifth Fibonacci number, which is 5. The answer is 55, which appears in the Fibonacci sequence. This property always holds.

Consider any two consecutive Lucas numbers. Say, 1 and 3. Square each of them and add the results. The answer is 10. This equals five times 2. The number 2 is the third Fibonacci number. Consider 3 and 4. Square each of them and add the results. The answer is 25. This equals five times 5. The number 5 is the fifth Fibonacci number. Consider 4 and 7. Square each of them and add the results. The answer is 65. This equals five times 13, and 13 is the seventh Fibonacci number. This property always holds.

The following are some lesser-known properties.

Consider the first two Lucas numbers. These are 1 and 3. Their product is 3. This is 1 less than the third Lucas number, 4. Consider the second and third Lucas numbers. These are 3 and 4. Their product is 12, which is 1 more than the fifth Lucas number, 11. This property always holds. The product of two consecutive Lucas numbers will be either 1 less or 1 more than a Lucas number.

Similarly, the square of any Lucas number, plus or minus 2, is always another Lucas number. For example, the square of 1 is 1, and 1 plus 2 is 3, which is a Lucas number. The square of, say, 7, is 49, and when 2 is subtracted from 49 we obtain 47, which is a Lucas number.

Sum the squares of four consecutive Lucas numbers—say, $1^2 + 3^2 + 4^2 + 7^2$. The answer, 75, is just 1 less than a Lucas number; and, as we know, 1 is the first Lucas number. Suppose one wants to find the sum of $3^2 + 4^2 + 7^2 + 11^2$. The answer is 195, which is 4 less than another Lucas number. And 4 is the third Lucas number. Suppose one finds the sum of $4^2 + 7^2 + 11^2 + 18^2$. The answer now is 510, which is 11 less than a Lucas number. If we check, we will find that 11 is the fifth Lucas number.

Similarly, the sum of the squares of 7, 11, 18, and 29 is 1,335, which is 29 less than a Lucas number, and, if we check, we'll see that 29 is the seventh Lucas number. In each successive case, beginning with the first four Lucas numbers, the sum of the squares of four consecutive Lucas numbers is less than the next Lucas number by 1, 4, 11, 29, 76, 199, and so on, each successive term being every second Lucas number.

Suppose one needs to sum the first n Lucas numbers. Check the $n + 2$ Lucas number and subtract 3 from it. That will be the answer to the sum. For example, to find the sum of the first six Lucas numbers, look up the eighth Lucas number, which is 47, and subtract 3 from it, which gives you 44. Therefore, the sum of the first six Lucas numbers is 44.

To calculate the sum of the squares of the first n Lucas numbers, calculate L_n multiplied by L_{n+1}, and subtract 2 from your result. For instance, to calculate the sum of the squares of the first four Lucas numbers, multiply the fourth Lucas number, 7, by the fifth Lucas number, which is 11. The answer is 77. Subtract 2 from this and we obtain 75. We find that $1^2 + 3^2 + 4^2 + 7^2 = 75$. To find the sum of the squares of the first eight Lucas numbers, we multiply the eighth Lucas number by the ninth Lucas number. We find that 47 multiplied by 76 equals 3572. Subtract 2 from this and we obtain 3570. We find that $1^2 + 3^2 + 4^2 + 7^2 + 11^2 + 18^2 + 29^2 + 47^2 = 3570$.

Consider the square of every second number in the Lucas sequence, beginning with 4. Subtract the square of the Lucas number that appears two places above that Lucas number. The result will be five times a Fibonacci number, beginning with 3. Using this process, we obtain the following numbers:

$$(4^2 - 1^2) = 5 \cdot 3$$
$$(7^2 - 3^2) = 5 \cdot 8$$
$$(11^2 - 4^2) = 5 \cdot 21$$
$$(18^2 - 7^2) = 5 \cdot 21$$
$$(29^2 - 11^2) = 5 \cdot 144$$
$$(47^2 - 18^2) = 5 \cdot 377$$
$$(76^2 - 29^2) = 5 \cdot 987$$

The following equation also holds:

$$Fn = \frac{(\mathrm{Ln}-1)+(\mathrm{Ln}+1)}{5}.$$

Thus, the twelfth Lucas number, 322, plus the fourteenth Lucas number, 843, is 1,165. This equals five times 233, and 233 is the thirteenth Fibonacci number. The fifteenth Lucas number, 1,364, plus the seventeenth Lucas number, 3,571, is 4,935. This equals five times 987, and 987 is the sixteenth Fibonacci number. And so on.

The sides of a Pythagorean triangle can be obtained if the following procedure is performed on any four consecutive numbers from the Lucas sequence. Consider the first four terms, 1, 3, 4, and 7.

Multiply the two inside numbers together ($3 \cdot 4 = 12$) and double that result, which equals 24. Multiply the two outside numbers together, which gives you 7. We find that 7 and 24 are two legs of a Pythagorean triangle. The hypotenuse will be 25. Thus the right triangle will be $7^2 + 24^2 = 25^2$.

The hypotenuse in such a triangle will always be five times a Fibonacci number. For example, in the case just mentioned, the hypotenuse is five times

5. We find that 5 is a Fibonacci number. The *area* will be the product of the four relevant Lucas numbers that generate the right triangle. In this case, the area of the 7, 24, 25 right triangle is (7 · 24)/2, or 84. We find that 1 · 3 · 4 · 7 = 84. The *perimeter* will be equal to twice the product of the last two of the four relevant consecutive Lucas numbers. The perimeter of the 7, 24, 25 right triangle is 7 + 24 + 25 = 56. The product of the last two relevant consecutive Lucas numbers is 28 (4 · 7). Twice 28 is 56.

As with the Fibonacci sequence, there is a formula that gives the *exact* *n*th number of the Lucas sequence. The formula is:

$$Ln = \left(\frac{1+\sqrt{5}}{2}\right)^n + \left(\frac{1-\sqrt{5}}{2}\right)^n.$$

However, just as with the Fibonacci numbers, there is a simpler method to find the *n*th Lucas number. Simply raise phi to the power of *n* and round to the nearest integer. Thus, to find, say, the twentieth Lucas number, raise phi (1.61803398 . . .) to the power of twenty, obtaining 15,126.99993 . . . , and round this off to 15,127, which you will find is the twentieth Lucas number. To find, say, the twenty-seventh Lucas number, raise phi to the power of twenty-seven, which is 439,203.98717, and round off to the nearest integer, 439,204. We find that 439,204 is the twenty-seventh Lucas number.

There is only one cube in the Lucas sequence: the number 1. There are only two square numbers in the Lucas sequence: 1 and 4. The only numbers that appear in both the Lucas and Fibonacci sequences are 1 and 3.

If you go up the powers of ten, you see a pattern in the distribution of the number of digits in the Lucas sequence. There is 1 digit in the first Lucas number. There are 3 digits in the tenth Lucas number; 21 digits in the one-hundredth Lucas number; 209 in the thousandth Lucas number; 2,090 in the ten-thousandth Lucas number; 20,899 in the one-hundredth thousandth Lucas number; and 208,988 digits in the one millionth Lucas number. The ten-millionth Lucas number contains 2,089,877 digits. These numbers converge on the number 208,987,640,249,978,733,769, which corresponds to the base ten logarithm of phi, which is 0.20898764 . . . .

Suppose you are given a number and asked if it is a Lucas number. How can you check if it is? Suppose, for example, you are given 521. Is it a Lucas number?

Here's a simple way of checking. First square the number you have been given. We find that $521^2 = 271{,}441$. Multiply this by 5. We obtain 1,357,205. Now add 20 to 1,357,205, obtaining 1,357,225, and subtract 20 from 1,357,205, obtaining1,357,185. Obtain the square root of both of these numbers, that is, the square root of 1,357,225 and the square root of 1,357,185. If either square root is an exact integer, then 521 is a Lucas number. If neither of the square roots is an integer, then 521 is not a Lucas number.

In this example we find that the square root of 1,357,225 is exactly 1,165 (an integer), and we find that the square root of 1,357,185 is 1,164.9828 . . . . Since the square root of 1,357,225 is an integer, the test tells us that 521 is indeed a Lucas number.

The test can be expressed in mathematical language as follows: let x be the number you are testing to see if it is a Lucas number. If $5x^2 \pm 20 = y^2$, where y is an integer, then x is a Lucas number.

The test has a bonus. When you obtain the value of y, and if it is an integer, do the following: Subtract x from y and divide by 2. Then add x to y and divide by 2. The first answer will be the Lucas number immediately preceding x in the Lucas sequence. The second result will be the Lucas number immediately following x in the Lucas sequence.

An example will make this procedure clear. In our example we were asked to check if 521 is a Lucas number. The test told us that y^2 is equal to 1,357,225 or 1,357,185. When y^2 equals 1,357,225, we find that y equals to 1,165, which is an integer. Therefore, 521 is a Lucas number.

Consider the number 1,165, which we found equals y in our check. Subtract 521 from 1,165, obtaining 644, and divide by 2. The answer is 322. This is the Lucas number that immediately precedes 521 in the Lucas sequence.

Consider 1,165 again. Now add 521 to 1,165, obtaining 1,686. Divide this by 2, obtaining 843. We find that 843 is the Lucas number that immediately follows 521 in the Lucas sequence.

Recreational mathematicians are constantly searching for connections between various and apparently unrelated mathematical entities. So it was inevitable that some number fans would sooner or later investigate if Lucas numbers are in any way related to magic squares. Such connections—if they exist—would strengthen the mathematician's view that all of mathematics is connected in some way. (We discussed magic squares in chapter 1.)

Lucas numbers are connected with magic squares in a surprising way. The normal three-by-three magic square with the digits from 1 to 9 is shown in figure 6.1:

8	1	6
3	5	7
4	9	2

Figure 6.1

The sum of the numbers in each of the three horizontal rows, in each of the three vertical columns, and in each of the two diagonal rows is 15.

Consider any nine consecutive Lucas numbers. Suppose you choose 7, 11, 18, 29, 47, 76, 123, 199, and 322. Place the Lucas numbers in a horizontal row and number them from 1 to 9 as follows:

7	11	18	29	47	76	123	199	322
1	2	3	4	5	6	7	8	9

Now substitute each Lucas number with its corresponding number in the magic square shown in figure 6.1. For example, where the number 1 appears in the magic square shown in figure 6.1, substitute it with the number 7. Where the number 2 appears in the magic square, substitute it with the number 11. And so on. You will then form the square that is shown in figure 6.2.

199	7	76
18	47	123
29	322	11

Figure 6.2

Consider now the products of each of the horizontal rows and the product of each of the vertical columns. You will find that:

$$199 \cdot 7 \cdot 76 = 105{,}868$$
$$18 \cdot 47 \cdot 123 = 104{,}058$$
$$29 \cdot 322 \cdot 11 = 102{,}718$$

The total sum of the three products is 312,644.

The products of the vertical columns are:

$$199 \cdot 18 \cdot 29 = 103{,}878$$
$$7 \cdot 47 \cdot 322 = 105{,}938$$
$$76 \cdot 123 \cdot 11 = 102{,}828$$

The total sum of the three products is again 312,644.

I sent Dr. Cong these observations about the Lucas sequence and the Fibonacci sequence before I sent them to the publisher. I asked him if he would furnish me with some interesting facts about the Fibonacci and Lucas numbers, and he did not disappoint.

Hello, my friend,

Thank you for sending your thoughts on the Lucas numbers. Here is some more information on this topic that your readers might enjoy.

Consider the equation $x^2 - xy - 5 = y^2$. It has solutions in integers only if x is a Lucas number that has an *even* subscript [the subscript is the number's position in the Lucas sequence] greater than 1, and y is the preceding Lucas number. For example, if the Lucas number's subscript is 6, then we have L_6, which equals 18, and L_5, which equals 11. Therefore in the above equation x equals 18 and y equals 11. The above equation then goes as follows:

$$18^2 - 18 \cdot 11 - 5 = 11^2.$$

It can be proved that every Lucas number divides some Fibonacci number. It can also be proved that no Fibonacci number greater than or equal to 5 divides any Lucas number.

The only Lucas number that is 1 less than a square is 3, and the only Lucas number that is 1 less than a cube is 7.

Here are a few curiosities involving both Fibonacci and Lucas numbers. The first two digits of the Fibonacci sequence, when concatenated, form a number (11) that appears in the Lucas sequence. The first two digits of the Lucas sequence, when concatenated, form a number (13) that appears in the Fibonacci sequence.

Consider the first three numbers of the Fibonacci sequence (1, 1, 2) and the Lucas sequence (1, 3, 4). Construct two three-digit numbers from these digits as follows: 112 and 134. We find that $134 - 112 = 22 = 2 \cdot 11$ and $134 + 112 = 246 = 2 \cdot 123$. Note that 11 is the 5th Lucas number and that 123 is the (5 + 5)th Lucas number.

The (–2 + 9)th Lucas number is 29, and the ($3 \cdot 2 \cdot 2$)th Lucas number equals 322. The (1 + 1 + 4 + 9 + 8 + 5 + 1)th Lucas number equals 1,149,851.

The digits 4 and 7 and the number 47 appear in the Lucas sequence. The following is a curious equation:

$$1{,}347 = 13 - 47 - 11 - 18 + 29 \cdot 47 - 76 + 123$$

(Note that the Lucas digits appear in order on both sides of the equation.)

Incidentally, Lucas was born in France. Using the simple code where A = 1, B = 2, C = 3, and so on, the sum of the letters in the word FRANCE is 47. Curiously, 47 is a Lucas number.

Finally, here is how the Lucas numbers 1, 3, 4, 7, 11, 18, 29, 47, and 76 are connected to one of the most beloved presidents of the United States, John F. Kennedy.

John F. Kennedy was the 1st United States president born in the twentieth century. He was born on the 3rd day of the week. His first name contains 4 letters and his surname contains 7 letters. He was killed in the 11th month of the year. His reputed assassin was born on the 18th day of October. John F. Kennedy was born on the 29th day of the month. John F. Kennedy was in his 47th year when he was assassinated. The police officer, Nick McDonald, who arrested Lee Harvey Oswald in the Texas Theatre in Dallas on November 22, 1963, died in hospital on January 27, 2005. He was 76 years old.

From your (not-so-old) friend,

Dr. Cong

CHAPTER 7

THE IRRATIONAL NUMBER PHI

The number phi (pronounced "fi" or "fee") is one of the three basic constants in mathematics. The other two are π, which equals 3.14159 . . . , and the number known as e, which equals 2.71828 . . . . We will discuss these other two constants later in this book.

Phi equals 1.6180339 . . . and is often referred to as the *golden ratio*, or the *divine proportion*. It crops up in many unexpected places in the world of mathematics. It is not known who first discovered phi, or how that person happened to discover this extraordinary number.

Phi is sometimes represented by the symbol Φ. The reciprocal of phi (1/phi) is sometimes represented as φ, which equals 0.6180339 . . . . To avoid confusion, we will let Φ in this chapter equal the number 1.6180339 . . . and we will let φ equal 0.618033988 . . . .

The number Φ has many curious properties. For instance, Φ is the only positive number such that Φ^2 equals $\Phi + 1$. Also, Φ is the only positive number such that 1 divided by Φ equals $\Phi - 1$. Also $1/\Phi$ equals 0.6180339 . . . , which equals φ. This means that $\Phi \cdot \varphi = 1$. The decimal expansion of φ is the same as that of Φ.

The negative form of φ is also possible, and it is calculated as follows:

$$-\Phi = \frac{1-\sqrt{5}}{2} = -0.6180339\ .\ .\ .\ .$$

However, it is the *positive* ratio, 0.61803398 . . . that is normally used in mathematics

One of Φ's best-known appearances in mathematics is in the Fibonacci and Lucas sequences, which we discussed in earlier chapters. The ratio between two consecutive numbers in both of these sequences (and indeed all additive sequences, no matter what their first two numbers may be) converge on 1.618033988, which equals Φ.

To see how adjacent Fibonacci numbers approach the value of Φ, consider the first few terms of the Fibonacci sequence: 1, 1, 2, 3, 5, 8, 13, 21, 34, 55, 89 . . . . The ratio 34/21, for example, equals 1.615 . . . , and the ratio 55/34 equals 1.61764705 . . . . As you go farther out into the Fibonacci sequence, the ratio of any two consecutive numbers (where the first of the two numbers is the largest) gets closer and closer to Φ but never quite equals it. Why? Phi is an irrational number, so it is not possible for a fraction, a/b, where a and b are two integers, to equal Φ.

If the two numbers that are chosen are such that the first one is smaller than the second, the ratio converges on 0.61803398 . . . or φ. For example, if the two numbers chosen are 89 and 144, the ratio between them is 89/144, or 0.61805555 . . . If the two chosen numbers are 144/233, the ratio between them is 0.61802575 . . . . And so on.

These properties of the Fibonacci sequence also apply to the Lucas numbers. The reader may recall that the first few terms of the Lucas sequence are 1, 3, 4, 5, 7, 11, 18, 29, 47, 76, 123, 199, 322, 521 . . . . The ratio of the fraction 47/29, for example, is 1.62068 . . . . And the ratio of the fraction 76/47 is 1.61702 . . . . As you go farther out the series, the ratio of two adjacent terms converge on Φ, if the first of the two terms is the largest. In contrast, if the first of the two terms is the smaller of the two, as in, say, 322/521, we find that this fraction equals 0.61804222 . . . . As we go farther out the sequence, the ratio of the relevant fractions converge on 0.618033988 . . . , which is φ.

The geometrical value of Φ most simply arises as follows. Draw a horizontal line and divide it into two parts, A and B, so that the length of $A + B$ is to A as A is to B. (See figure 7.1.)

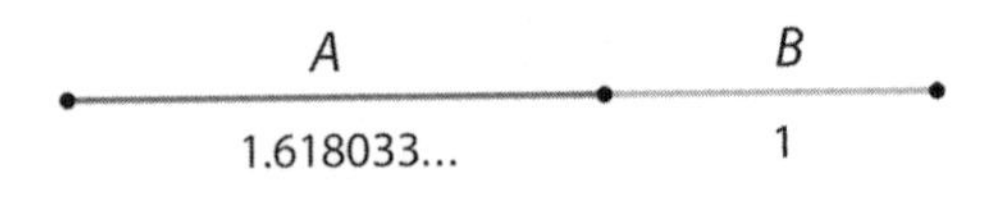

A+*B* is to *A* as *A* is to *B*

Figure 7.1

Let the distance B equal 1. The following identity can now be written:

$$\frac{A+1}{A} = \frac{A}{1}$$

This identity equals the quadratic equation:

$$A^2 - A = 1 = 0$$

Solving this quadratic gives A the positive value of

$$\Phi = \frac{1+\sqrt{5}}{2} = 1.618033988749894\ldots.$$

Therefore, assuming the length of the segment B in the horizontal line above is 1, the length of the entire line $A + B$ is divided in the golden ratio: 1.61803398 . . . to 1.

Because Φ is an irrational number, its decimal expansion goes on forever. Computer programmers have devised programs to calculate millions of decimals of Φ. Here is Φ to thirty decimals:

1.618033988749894848204586834365

The number Φ crops up a great deal in plane geometry. Consider the pentagon, for instance. (See figure 7.2.) If each side of the pentagon is equal to 1, then the length of any of the interior diagonals, such as *AB*, or *CD*, is equal to 1.618033987 . . . .

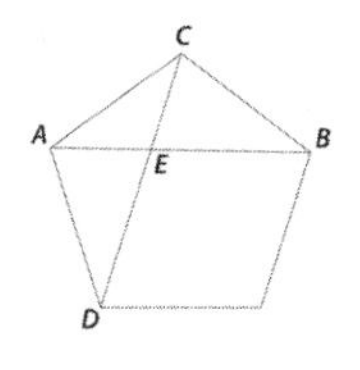

Figure 7.2

There are numerous appearances of Φ in other geometrical figures as well. Suppose you construct a rectangle so that its long side and its short side are in the ratio of Φ. This figure is known as a *golden rectangle*. (See figure 7.3.)

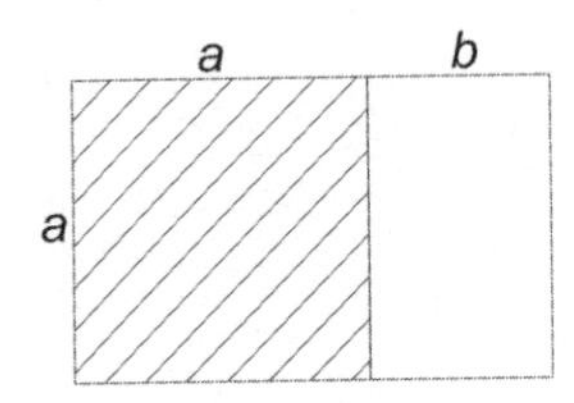

Figure 7.3

If you cut off a square whose side is equal to the shortest side of this rectangle, the rectangle remaining is also a golden rectangle. For example, suppose the original rectangle measures 40.4508497 units in length and 25 units in width. If you cut off a square measuring 25 by 25, the resulting rectangle that's left will measure 15.4508497 units in length and 25 units in width. The ratio between these two sides is 1.618033988 . . . , which equals Φ. This result is one of the many reasons Φ acquired the name *golden ratio*.

The reason it acquired this name is that the ratio was considered special. If we cut off a square from a rectangle whose long and short sides are NOT in the golden ratio, the long and short sides of the rectangle that remains will NOT be in the ratio of the original rectangle. The remaining rectangle will only retain its original proportion after the square piece is cut from it if, and only if, the original rectangle has long and short sides in the proportion of the golden ratio. It is this unique property of the ratio that makes it special among the infinity of ratios that is possible.

In 2007, Gary W. Adamson (an amateur mathematician living in the United States) found that Φ can be expressed as the sum of the infinite product of the reciprocals of every second number that appears in the Fibonacci sequence, beginning with 1. (See figure 7.4.)

$$\Phi = \frac{1}{1} + \frac{1}{2} + \frac{1}{2 \cdot 5} + \frac{1}{5 \cdot 13} + \frac{1}{13 \cdot 34} + \frac{1}{34 \cdot 89} + \frac{1}{89 \cdot 233} + \ldots$$

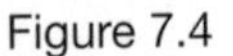

Figure 7.4

This series illustrates the intimate link between Φ and the Fibonacci numbers.

The fundamental nature of Φ is revealed when it is expressed as an infinite continued fraction:

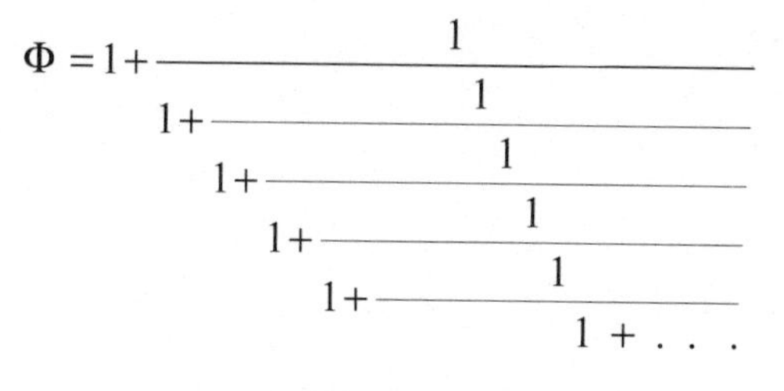

$$\Phi = 1 + \cfrac{1}{1 + \cfrac{1}{1 + \cfrac{1}{1 + \cfrac{1}{1 + \cfrac{1}{1 + \ldots}}}}}$$

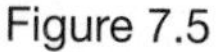

Figure 7.5

This is the only continued fraction in all of mathematics that consists entirely of 1s. This fact alone should tell us that Φ is a special number.

This infinite continued fraction, though beautifully simple, converges on the true value of Φ extremely slowly. In fact, Φ is often described by mathematicians as the worst-performing continued fraction because it is the slowest continued fraction to converge on its designated value. Of course, the farther you calculate the infinite continued fraction, the closer you get to the true value of Φ, but you will never quite attain that *exact* value.

Every additive series beginning with any two numbers strives to reach a ratio equal to Φ. For example, suppose we start a series with 3 and 7. The series then is 3, 7, 10, 17, 27, 44, 71, 115, 186 . . . . The ratio between any two adjacent numbers in the series approaches Φ. For example, 7/3 equals 2.3333333 . . . . But 10/7 equals 1.4285714 . . . . We see that 17/10 equals 1.7, and 27/17 equals 1.588235294 . . . . The farther we go out the series, the closer the fractions converge on Φ.

If we form fractions from the above series by placing the smaller number as the numerator, we obtain the fractions 3/7, 7/10, 10/17, and so on. As we go out farther into the series, we find that these fractions also converge on 0.618033987 . . . , which equals φ.

The ratio between successive terms of only *one* additive series actually equals Φ; the series of successive powers of Φ. (See figure 7.6.)

$$1\Phi = \Phi^1$$
$$1\Phi + 1 = \Phi^2$$
$$2\Phi + 1 = \Phi^3$$
$$3\Phi + 2 = \Phi^4$$
$$5\Phi + 3 = \Phi^5$$
$$8\Phi + 5 = \Phi^6$$
$$13\Phi + 8 = \Phi^7$$
$$\ldots$$

Figure 7.6

This series reveals its intimate connection with the Fibonacci sequence. The left-hand column and the middle column in figure 7.6 reveal the Fibonacci sequence. The right-hand column gives the successive powers of Φ, beginning with Φ to the power of 1. Note that the values on the left-hand side in figure 7.6 are equal to the successive powers of Φ (beginning with the third power). These values are equal to the sum of the two previous powers. For example, $(1\Phi) + (1\Phi + 1) = 2\Phi + 1$; $(1\Phi + 1) + (2\Phi + 1) = 3\Phi + 2$; $(2\Phi + 1) + (3\Phi + 2) = 5\Phi + 3$; and so on.

Figure 7.7 shows another beautiful pattern linking the Lucas numbers and the Fibonacci numbers. The Fibonacci numbers appear on the extreme left-hand side in this pattern. The Lucas numbers are the numbers immediately to the right of the $\sqrt{5}$ in each successive row of figure 7.7. On the right-hand side of the equation signs are the successive powers of Φ.

$$\frac{1 \cdot \sqrt{5} + 1}{2} = \Phi^1$$

$$\frac{1 \cdot \sqrt{5} + 3}{2} = \Phi^2$$

$$\frac{2 \cdot \sqrt{5} + 4}{2} = \Phi^3$$

$$\frac{3 \cdot \sqrt{5} + 7}{2} = \Phi^4$$

$$\frac{5 \cdot \sqrt{5} + 11}{2} = \Phi^5$$

$$\ldots$$

Figure 7.7

It is unfortunate that a cult following around the world has arisen in connection to the number Φ. The members of this cult are continuously exaggerating the appearances of Φ in nature. Consequently, much nonsense has been written about this number over the decades. Today much of this is repeated on various sites on the Internet.

For example, there are those who argue that Φ shows up in the anatomy of a human being. Arguments have been made that the height of the average man or woman compared to the height of that individual's naval is Φ, measured from the floor. These claims are false.

There are also claims made that the most pleasing rectangle to the human eye occurs when the ratio between the width and length are equal to Φ. This claim originated with Gustav Theodor Fechner (1801–1887), the German philosopher and physicist. He was reported to have carried out experiments with a large number of participants. According to the results of the experiments, the majority of these participants preferred rectangles whose width and length were close to the ratio known as Φ. However, it later emerged that Fechner cooked the results and discarded those preferences that did not include Φ as the preferred ratio. Today, most mathematicians refute the results of Fechner's experiments.[1]

In addition to articles on the Internet, numerous articles in magazines and journals refer to Φ's repeated appearances in natural phenomena. Readers should treat many of the numerous claims made in these articles with much skepticism. One claim that has been made regularly is that Φ appears in the shape of spirals in sea shells.[2] By simply taking a relatively large number of measurements, this claim has been proven to be false.[3] The claim has also been made that Φ appears in the spirals of galaxies;[4] this claim is also false.[5]

If I was teaching students about Φ, I would try to convey the sheer wonder of this special number to them. I would do this by pointing out to them the obvious fact that the value of Φ lies between one and two. I would then point out that Φ, and the number *e*, which equals 2.71828 . . . and pi, which equals 3.14159 . . . , are all less than four. I would stress that this remarkable fact is strange for the following reason: There are an infinite number of integers; yet for some reason unknown to the human mind, three of the most ubiquitous numbers in mathematics are all less than four. Why is this the case?

No one knows for sure! It is one of the great mysteries of mathematics. But I would convey to the students that this curious fact simplifies mathematical calculations and enables human beings to learn at least some of the secrets of nature.

I would also point out that noticing a curious coincidence like this can sometimes lead to other deep questions of how the universe operates.

When a curious coincidence such as this is noticed, it often leads students to ask deep questions about numbers and where they crop up in nature. For example, a bright student may ask why the universe apparently contains only three spatial dimensions. Why three dimensions? Why not four, or one hundred, or one hundred million? Is there something special about the number three?

Probably not!

But the fact that the universe appears to have only three dimensions has enormous consequences for all life.

In our universe of three spatial dimensions, the force of gravity between two large celestial bodies, such as stars or planets, depends on the distance between them squared. It is known that if a universe contained only two dimensions, the force of gravity between two large celestial objects in space depends on just the distance between the two objects. In a universe containing four dimensions, the force of gravity between two celestial objects depends on the distance between then cubed.

What are the consequences of these facts?

Well, in a two-dimensional universe, the strength of gravity would be too strong to allow planets to orbit stars. Consequently, solar systems could not form and life—as we know it—could not exist. In a four-dimensional universe, gravity would be too weak and once again solar systems could not form. It is only in a three-dimensional universe that gravity would be the correct strength for stars, planets, and solar systems to form, and so allow the conditions for life to form and evolve.

So it appears that life only exists because there are three spatial dimensions.[6]

I would convey this extraordinary fact to students to bring home to them the joy of learning about numbers and how this knowledge can then lead to other deeper insights about the amazing universe we inhabit.

Platonist mathematicians—who believe that we discover mathematical truths—have long considered numbers such as Φ, *e*, and pi as special. Anti-Platonist mathematicians—who believe we invent mathematics—take the opposite view. Those anti-Platonist mathematicians who believe that Φ is not a special ratio use various arguments to support their point of view. One argument that has been frequently used involves the heptagon, which is a seven-sided polygon. (See figure 7.8.)

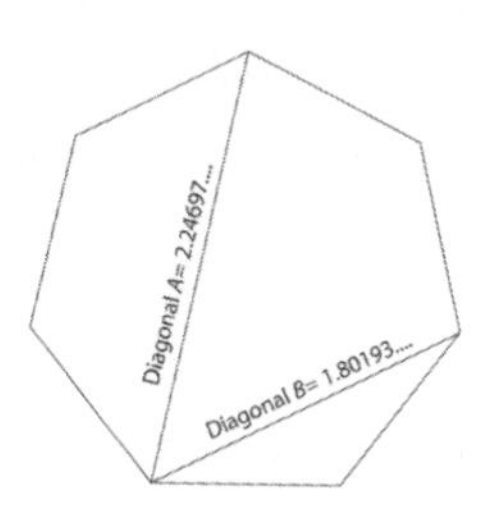

Figure 7.8

Consider the heptagon with a side equal to 1. There are two different-length diagonals inside a heptagon: the longest diagonal is 1 is 2.24697 . . . (let's call that length A), and the shortest diagonal is 1.80193 . . . (let's call that length B). Various beautiful relationships exist between these two lengths. For example, $A - 1 = A/B$; $1/A = A + B$; $AB = A + B$; $A^2 = A + B + 1$; $B^2 = A + 1$, and so on. So even in the heptagon interesting ratios are found to exist.

These examples, and other similar ones, have led many anti-Platonist mathematicians to argue that Φ and the number curiosities involving Φ are nothing out of the ordinary.

Perhaps they are correct. Who knows for certain?

However, as I've mentioned before, mathematical Platonists believe in the independent reality of mathematics. They believe that Φ is a fundamental constant that has always existed. They believe, for example, that the hundredth decimal digit of Φ is 4, and that it always has been 4, long before human beings walked the Earth. Furthermore, they believe that the hundredth decimal digit of Φ will always be 4 long after the universe has disappeared. These mathematicians also believe, for instance, that Φ plus 1 is equal to Φ multiplied by itself, not because we human beings deem that result to be so, but because it *is* so! That, they insist, is the nature of mathematical reality.

Galileo wrote that the laws of nature are written in the language of mathematics. Many mathematicians sincerely believe this to be the case. But no one knows *why* it is the case. It is a tremendous mystery that human beings will probably never solve. It is something that we must accept as a given. The numbers e, π, 1, 0, i, and $-i$ are numbers that are ubiquitous in mathematics. Whatever their origin, they are, in the opinion of many of the greatest mathematical minds, significant numbers that keep cropping up in various areas of mathematics and physics.

The late mathematics and science writer Martin Gardner was a mathematical Platonist. He believed that mathematical truths exist "out there" and are eternal. Gardner also believed that the universe has a mathematical structure. He once wrote: “Mathematics is not only real, but it is the only reality. That is that the entire universe is made of matter, obviously. And matter is made of particles. It’s made of electrons and neutrons and protons. So the entire universe is made out of particles. Now what are the particles made out of? They’re not made out of anything. The only thing you can say about the reality of an electron is to cite its mathematical properties. So there’s a sense in which matter has completely dissolved and what is left is just a mathematical structure.”[7]

In other words, Gardner was saying that when one drills down to the level of the smallest particles of nature, matter as we know it appears to vanish and what remains are the mathematical properties that describe matter.

To see why many mathematicians and physicists believe that mathematical reality is “out there” and is not just a mere human invention, let me give just one example. At the beginning of the twentieth century, physicists were aware that electrons with positive energy exist in nature. No electrons with negative energy had been found, and there was no reason to believe that they might exist. The idea that an electron with negative energy could exist was completely dismissed as being against common sense.

However, in 1928, the great English physicist Paul Dirac (1902–1984) stated that an electron with both positive energy and negative had to exist in nature. He based this prediction purely on the results of a mathematical equation. That equation involved the square root of a specific quantity. The laws of mathematics tell us that the square root of a quantity involves two solutions: one solution is positive and one is negative. For example, the square root of 25 is + 5 or –5. The two solutions in Dirac’s equation meant that two types of electron are possible in nature: one electron has positive energy, and the second electron has negative energy. This was completely contrary to the scientific viewpoint at that time, but this is what the laws of mathematics were saying. Dirac placed his faith in the power of mathematical reality and declared that electrons with negative energy had to exist.[8]

Lo and behold, on August 2, 1932, an American physicist, Carl David Anderson (1905–1991), discovered an electron with negative energy, a feat that earned Anderson the Nobel Prize for Physics in 1936. The electron with negative energy was eventually named the *positron*, and its existence was the first confirmation of antimatter in the universe. The unearthing of this deep secret of nature confirmed Dirac’s theoretical prediction and justified his belief in the power of mathematical equations.

There have been many other examples where the results of mathematical equations mysteriously predicted the existence of subatomic particles. That there is such a deep connection between pure mathematics and physics is astonishing. Many mathematicians can only conclude from this that the workings of the universe mysteriously follow mathematical laws. *Why* this is so will probably never be known.

Mathematical Platonists around the world believe that mathematical truths are *discovered* and are not *invented*. Many physicists also share this viewpoint. The German physicist Heinrich Rudolf Hertz (1857–1894) said: "One cannot escape the feeling that these mathematical formulae have an independent existence and an intelligence of their own, that they are wiser than we are, wiser even than their discoverers, that we get more out of them than we originally put in to them."[9]

Returning momentarily to Paul Dirac, he once said that "it is my fundamental belief that the laws of nature should be expressed in beautiful equations."[10] That viewpoint is the core belief of many great scientists, mathematicians, and philosophers. It astonishes many of the greatest minds on Earth that such numbers as e and π (and the so-called imaginary numbers, i or $-i$) arise in formulae that describe various natural phenomena.

For example, π appears in Einstein's *field equation*. This equation relates the curvature of space-time with energy sources, and it is the basic principle of relativity theory.[11] The fact that π appears in this equation is looked upon with awe and wonder by many physicists and mathematicians. Physicists find that these numbers, especially π, e, and the imaginary i, appear to be essential ingredients of the mathematical structure that governs the laws of nature, or as a way of describing the laws of the universe, depending on their point of view as to the nature of mathematics.

Physicists, mathematicians, and philosophers often wonder why some of the basic rules of nature can be expressed so elegantly by mathematical formulae. For example, Albert Einstein (1879–1955) discovered that energy and matter are two different forms of the same thing. This fact can be expressed in this simple-looking equation: $E = mc^2$. In this equation, E represents energy, m represents mass, and c^2 represents the speed of light squared.

Why should such a fundamental law of the universe be capable of being expressed with just a few mathematical symbols in a one-line equation such as this? Reality does not have to be this way. It seems that things could have been far different. Perhaps we could have inhabited a universe where ten million characters would have to be written to explain such a basic law of nature. Luckily for us, we don't live in such a universe. It is surely a great mystery that intelligent beings—that are part of the universe for a relatively short period of time—are capable of describing a basic law of the universe such as this with just a few characters. We are—it seems—the universe's way of it becoming aware of itself.

Getting back to the number Φ, it is remarkable that this number can also be expressed as follows:

$$\phi = \sqrt{1+\sqrt{1+\sqrt{1+\sqrt{1+\ldots}}}}$$

The proof of this identity is remarkably simple. Let N equal the total sum on the right-hand side of the equation. Let x equal the sum of the right-hand side of the equation after the first 1. We can now write:

$$N = \sqrt{1+x}$$

Squaring both sides of the equation, we obtain:

$$N^2 = 1 + x$$

or

$$N^2 - 1 - x = 0.$$

This is the quadratic equation that has a root equal to Φ, and a second root equal to –φ.

Φ can also be simply expressed using four 5s (three of which are 0.5) as follows:

$$5^{.5} \cdot .5 + .5 = \Phi.$$

In other words, this equation states that 5 is first raised to the power of 0.5; the result is then multiplied by 0.5; the result of that is then added to 0.5. The answer is 1.61803398 . . . , which is Φ. The equation follows immediately from the identity:

$$\Phi = \left(\frac{\sqrt{5}}{2}\right) + 0.5.$$

This is the usual formula for Φ with the terms rearranged.

The number Φ is the limit of the *n*th root of the *n*th Fibonacci number. For example, consider the thirteenth Fibonacci number. It is 233. The thirteenth root of 233 is the number that, when multiplied by itself thirteen times, equals 233. Mathematically, the thirteenth root of 233 is written as:

$$\sqrt[13]{233} = 1.52091275\ldots$$

Consider the thirtieth Fibonacci number. It equals 832,040. The thirtieth root of 832,040 is:

$$\sqrt[30]{832{,}040} = 1.57520883\ldots$$

This result is a little closer to the value of Φ.

The forty-fifth Fibonacci number is 1,134,903,170. The forty-fifth root of 1,134,903,170 is even closer to Φ.

$$\sqrt[45]{1{,}134{,}903{,}170} = 1.58935644\ldots$$

It can be proved that as *n* grows to infinity, the *n*th root of the *n*th Fibonacci number approaches the value of Φ.

There are numerous other curiosities concerning this unique irrational number. Of course, there are those who may ask what these curiosities tell us about our world. My answer to these questions is: They perhaps tell us nothing! Mathematics in itself does not *explain* our universe. Yes, mathematics can be used in physics to explain how some parts of the universe operate. That is truly marvelous. But that is not why mathematics exists. Mathematics exists in its own right. It may well be the *only* reality. To find explanations on how the world works, I suggest one should study physics.

The very interesting properties of numbers, including the special number known as Φ, strongly suggest, I believe, that there is sublime beauty found in numbers.

I asked my friend Dr. Cong if he would give me any curiosities concerning Φ that may be new to my readers. Dr. Cong spent much of his youth performing magic for a living and also spent some time giving paid lectures on recreational mathematics and lightning calculation. He sent me the following information by e-mail.

> Hi, Owen,
>
> Thank you for sending me your piece on the amazing number known as Φ. I found it most interesting. Incidentally, I am glad you made the argument that mathematics is not invented by intelligent minds, but is *discovered* by them. If there is intelligent life on other planets in this amazing universe, I have no doubt that they will also have discovered the special numbers Φ, *e* π, and the two so-called imaginary numbers, *i* and –*i*.
>
> You ask for a few curiosities concerning Φ. Here are some that your readers might like.
>
> Those readers who like trigonometry and have an interest in the notorious number of the beast, 666, may find the following expressions for Φ curious:
>
> $$\Phi = -2 \sin 666 \text{ degrees} = 1.618033988 \ldots.$$
>
> and
>
> $$\Phi = -2 \cos 6 \cdot 6 \cdot 6 \text{ degrees} = 1.618033988 \ldots.$$
>
> Consider the first four digits of Φ. These are 1618. Partition this number into the two parts 16 and 18. The product of these two numbers is 288. The sum of the proper divisors of 288 [the proper divisors of any number, *n*, are the divisors less than *n*] is 531. This number, 531, plus the reverse of the digits of this number, 135, equals 666. (Note that the divisors of 288 are 1, 2, 3, 4, 6, 8, 9, 12, 16, 18, 24, 32, 36, 48, 72, 96, 144, and 288. The sum of all these divisors is 819. The *proper* divisors of 288, however, are all the divisors *less* than 288. The sum of all the proper divisors of 288 is 531.)
>
> The following infinite series, in which the Fibonacci numbers appear in the denominators, sums to $\frac{1}{\Phi}$, which equals Φ, or 0.6180333988 . . . .
>
> $$\Phi = 1 + \frac{1}{1} - \frac{1}{1 \cdot 2} + \frac{1}{2 \cdot 3} - \frac{1}{3 \cdot 5} + \frac{1}{5 \cdot 8} - \frac{1}{8 \cdot 13} + \ldots$$

The number 5 is intimately connected to Φ. For example, 2 times Φ minus 1 equals the square root of 5. All one-digit numbers, except 5, appear at various places in the first 22 digits of Φ. The 23rd digit is 5. Thus, the last one-digit number to appear in the decimal expansion of Φ is 5. Note that the digits of 23 sum to 5. Phi appears in the pentagon, which is a five-sided figure. Phi and the Fibonacci sequence are intimately related. The fifth digit of the Fibonacci sequence is 5. Using the usual alphabet code where A = 1, B = 2, C = 3, and so on, the sum of the letters in the word PHI equals 33. Note that the number 33 appears in Φ beginning at the fifth decimal digit: 1.6180*33*988 . . . .

Erol Karazincir, a number enthusiast living in the United States, discovered that Φ can be simply expressed as follows:

$$\Phi = \sqrt{\left(\frac{5+\sqrt{5}}{5-\sqrt{5}}\right)}$$

Ask your readers, Owen, if they can prove this beautiful identity.

Here is a curious connection between Φ and the first five integers:

$$1/\Phi + 1/\Phi^2 = 1$$
$$\Phi + 1/\Phi^2 = 2$$
$$\Phi^2 + 1/\Phi^2 = 3$$
$$(\Phi + 1/\Phi^2)^2 = 4$$
$$(\Phi + 1/\Phi)^2 = 5$$

A surprising result is obtained if you sum the infinite series of the reciprocals of Φ. (The reciprocal of Φ is 1/1.618033988, which equals 0.618033988, which is φ.) This series equals $1/\Phi^1 + 1/\Phi^2 + 1/\Phi^3 + 1/\Phi^4 + \ldots$. Your readers may be surprised to learn that the series sums to Φ! Ask your readers if they can prove this.

Best regards,

Dr. Cong

SOLUTIONS

Here are the solutions to both of Dr. Cong's puzzles.

(1) Dr. Cong asked our readers to prove the following beautiful equation:

$$\Phi = \sqrt{\left(\frac{5+\sqrt{5}}{5-\sqrt{5}}\right)}$$

Here is one way of attacking the problem.

We write the following equations:

$$\Phi = \frac{\sqrt{5}+1}{2} \text{ or } 2\Phi = \sqrt{5}+1 \text{ or } 2\Phi\sqrt{5} = 5+\sqrt{5}$$

$$\Phi = \frac{\sqrt{5}-1}{2} \text{ or } 2\Phi = \sqrt{5}-1 \text{ or } 2\Phi\sqrt{5} = 5-\sqrt{5}$$

Therefore,

$$\frac{\left(2\Phi\sqrt{5}\right)}{2\Phi\sqrt{5}} = \frac{5+\sqrt{5}}{5-\sqrt{5}}.$$

Consider the fraction on the left-hand side of the above equation. All of the terms in the numerator and denominator cancel out, except Φ and φ, leaving

$$\frac{\Phi}{\Phi} = \Phi^2.$$

This equals the fraction

$$\frac{5+\sqrt{5}}{5-\sqrt{5}}.$$

Since

$$\Phi^2 = \frac{5+\sqrt{5}}{5-\sqrt{5}},$$

it follows that

$$\Phi = \sqrt{\left(\frac{5+\sqrt{5}}{5-\sqrt{5}}\right)}.$$

(2) We are asked to prove that the infinite series $1/\Phi^1 + 1/\Phi^2 + 1/\Phi^3 + 1/\Phi^4 + \ldots = \Phi$.

Rewrite the series we are asked to prove as

$$x = 1/\Phi^1 + 1/\Phi^2 + 1/\Phi^3 + 1/\Phi^4 + \ldots. \qquad \text{(Equation I)}$$

Note that the first two terms of this series equals 1. In other words,

$$1/\Phi^1 + 1/\Phi^2 = 1 \qquad \text{(Equation II)}$$

This allows us to rewrite Equation I as:

$$x = 1 + 1/\Phi^3 + 1/\Phi^4 + 1/\Phi^5 + 1/\Phi^6 \ldots. \qquad \text{(Equation III)}$$

Multiply across the equation by Φ^2 to obtain:

$$x \cdot \Phi^2 = \Phi^2 + 1/\Phi^1 + 1/\Phi^2 + 1/\Phi^3 + 1/\Phi^4 + \ldots. \qquad \text{(Equation IV)}$$

Note that the terms of the series (on the right-hand side of the equation) after Φ^2 are identical to the value of x as given in Equation I.

This allows us to write:

$$x \cdot \Phi^2 = \Phi^2 + x. \qquad \text{(Equation V)}$$

Dividing across the equation by x gives:

$$\Phi^2 = \frac{\Phi^2}{x} + 1. \qquad \text{(Equation VI)}$$

or

$$\Phi^2 - 1 = \frac{\Phi^2}{x}$$

or

$$\Phi = \frac{\Phi^2}{x}$$

or

$$\Phi x = \Phi^2.$$

Therefore,

$$x = \frac{\Phi^2}{\Phi}$$

or

$$x = \Phi.$$

CHAPTER 8

THE SQUARE ROOT OF –1

Mathematicians long ago realized that the one entity of mathematics they constantly used, the counting numbers, 1, 2, 3, 4, 5 . . . would have to be extended to provide solutions to problems that occur in the natural world. Thus the fractions, such as $\frac{1}{2}$, and $\frac{3}{4}$, were invented or discovered, depending on your point of view on the origin of mathematics. This was followed by extending the mathematical entities to include irrational numbers and negative numbers.

Why did mathematicians come up with these new mathematical entities? Well, it seems that mathematical reality forced them to accept that these entities are part and parcel of the mathematical landscape, and that we cannot easily solve mathematical problems, or indeed real-world problems, without them. So mathematicians accepted that these quantities somehow existed, either in their heads, or as abstractions in the physical universe.

That is the way things stayed until the middle of the fifteenth century. By the mid-1500s, mathematicians had reported that it was convenient sometimes to multiply a strange new type of number by itself, and get negative 1 (–1) as the result. (In the modern world, we call this apparently strange entity the number *i*.)

Mathematicians were aware, of course, that when dealing with the counting numbers, multiplying a number by itself resulted in a positive quantity. A positive number multiplied by a positive number gives another positive number. Similarly, a negative number multiplied by another negative number gives a positive number also. Thus a negative number multiplied by itself gives a positive number. But then in the mid-1500s this new idea started to spread among mathematicians that a number could somehow exist in which the square of the number equaled a negative quantity. This must have mystified many a mathematician when the idea was first suggested. The concept did, after all, appear to go against the basic laws of mathematics as they were understood at the time.

Why did mathematicians come up with the idea of the square root of –1? Well, the answer is that, once again, mathematical reality imposed itself on them. They found that there really was little choice but to accept that a number such as the square root of –1 exists in order to solve certain types of quadratic, cubic, and quartic equations. If mathematicians decided that there could not possibly be a number that equaled the square root of –1, then it meant that an innocent-looking equation such as $x^2 = -1$ had no solution. Most mathematicians believed that this consequence would be preposterous, and so slowly and gradually they came to recognize and accept that the square root of –1 does exist, and should be used in order to solve equations.

This is not the first time that mathematicians were confronted with the square root of –1. Nearly four thousand years ago, Babylonian mathematicians were able to solve simple quadratic equations when the solutions were positive numbers. They had a different method of solution than the one we use today, but nevertheless their method produced the right answers. Their number system was the most advanced in the world at the time, but they did not know what to do when they encountered negative numbers in their attempts to solve certain types of quadratic equations.

For example, given the quadratic equation $x^2 - x - 2 = 0$, one finds that there are two solutions: either $x = 2$ or $x = -1$. The Babylonians did not know what to make of this negative number, –1. How could there be a number such as –1? Surely such a number is less than zero, so how could such a number equal x in the given equation? The Babylonians decided to simply ignore negative quantities when they came up in their calculations. This held back any progress that might have been made in the advancement of mathematics. Why? Well, if the Babylonians accepted that negative numbers did indeed exist and were indispensable in mathematics, then perhaps mathematicians at the time would have been able to use them to solve more complicated equations such as the cubic and quartic.

Indeed, no progress was made in the next two thousand years in attempting to understand what these equations containing negative numbers meant.

However, in the 1500s, Italian mathematicians found that it was essential to assume the existence of a strange type of number: the square root of –1. This is the number we today call i. The Italian mathematicians realized that they could not solve certain types of cubic equations unless they used an entity that equaled the square root of –1.

They knew that every positive number has two square roots. For instance, the square roots of 25 are positive 5 and negative 5. When both positive 5 and negative 5 are squared, the result is positive 25. At the time, the Italian mathematicians were attempting to solve cubic equations. An example of a cubic equation is $x^3 + 2x^2 + 1 + 8 = 0$. Strange as it may seem, if we do not accept that the square root of –1 exists, we cannot solve this equation. In other words, we cannot find values of x that will make the equation correct. However, once we do accept that the square root of –1 exists, we find we can solve this equation, and an infinite number of similar equations. The three solutions to this equation (there are at most three solutions to cubic equations) are $x =$ –2.71618865 . . . , $x =$ 0.35809432 . . . + 1.67841352i, and $x =$ 0.35809432 . . . – 1.67841352i. In other words, if we put x equal to any one of these three values, the equation $x^3 + 2x^2 + 1 + 8$ is indeed equal to 0. If we make x equal to any other value, then $x^3 + 2x^2 + 1 + 8$ will not equal 0.

This development forced the Italian mathematicians to accept that the square root of –1(denoted as i) must exist. They accepted that it went against the common rule of mathematics that applied to everyday numbers at that time: a positive number multiplied by a positive number gives a positive number, and a negative number multiplied by a negative number gives a positive number. But now they had to accept that a number—or an entity—exists such that the square of this entity equals –1.

This entity, of course, could not be a number such as the counting numbers. Therefore, they described this number as an *imaginary* number, because they believed that the number merely existed in their imaginations. They were convinced that this number was less *real* than the ordinary counting numbers, although we now know that these so-called imaginaries are every bit as real as the ordinary numbers. They used the first letter of the word *imaginary* to represent this new number. So this new imaginary number was called i. To this day, mathematicians around the world still use i to represent the number that when squared equals –1.

Because the existence of the $\sqrt{-1}$ is assumed to be i, or $-i$, it is easy to work out the square roots of other negative numbers. For example, the square root of –3 is $3i$, because $3i^2$ equals 3 times –1, which equals –3. Similarly the square root of –4 is $2i$, because $(2i^2)$ equals 4 times –1, which equals –4. And so on.

Cubic equations are equations that have a quantity in them that is raised to the power of 3. The formula to solve quadratic equations (equations that contain a power of 2) was known, and so the search was on in Italy in the 1500s to see if a similar type of formula could be found that would solve the cubic equation. At the time, the Italian mathematician and philosopher Girolamo Cardano (1501–1576) was attempting to use a formula that was originally discovered by another mathematician named Niccolò Tartaglia (1499–1557). Tartaglia had discovered a formula that would solve cubic equations of the form $x^3 + bx^2 + cx + d = 0$. He eventually passed the details of this formula on to Cardano.[1]

Cardano played around with the formula and eventually discovered that every cubic equation of the form $x^3 + bx^2 + cx + d = 0$ could be transformed into the following one:

$$x = \sqrt[3]{\frac{q}{2} + \sqrt{\frac{q^2}{4} + \frac{p^3}{27}}} + \sqrt[3]{\frac{q}{2} - \sqrt{\frac{q^2}{4} + \frac{p^3}{27}}}$$

Cardano decided to use this formula to solve the following straightforward cubic equation:

$x^3 = 15x + 4.$

If we substitute 4 for x in the equation, we will get:

$64 = (15 \cdot 4) + 4$

$64 = 64$

So the solution is 4. In other words the value of x is 4.

Cardano knew, of course (without having to use his formula), that the solution to this equation is 4. He also knew that the formula he had derived to solve this form of cubic equation was correct. But the strange thing was that when he used the formula above to solve this cubic, he obtained the following solution:

$$x = \textit{cube root of} \sqrt{(2)+\sqrt{-121}} - \textit{cube root of} \sqrt{(-2)-\sqrt{(-121)}} = \textit{cube root of} \sqrt{(2)+\sqrt{-121}} + \textit{cube root of} \sqrt{(2)-\sqrt{(-121)}}.$$

Cardano did not know what to make of this. Why did the formula not give the correct solution to the equation, which is 4? Why was the formula returning an expression that included square roots of negative numbers? (The square root of –121 is negative.) What did the square root of –121 mean, and what was it doing in the assumed solution of the cubic?

Cardano was mystified. He finally had to accept that his formula gave correct results to cubic equations only in certain situations. He was resigned to the fact that the formula just did not appear to be able to solve all cubic equations. This was a deep mystery, because Cardano knew that the formula he was using was the *correct* formula passed to him by Tartaglia.

Cardano made no further progress on the matter. The situation remained unresolved for twenty years. Then in 1572 an Italian engineer, Rafael Bombelli (1526–1572), published a book titled *Algebra*.

Bombelli had never attended college but had received instruction in mathematics from another engineer by the name of Pier Francesco Clementi around the middle of the 1500s. Bombelli wrote his book for ordinary men and women who, like him, had not received a college education. His book was an extensive account of what was known about algebra at that time. More important, he explained in the book what negative numbers are and how to manipulate them. He was the first European to give such an account. When the great German mathematician Gottfried Leibniz read Bombelli's book, he praised Bombelli as an "outstanding master of the analytical art."[2]

Among other topics in his book, Bombelli states that imaginary numbers are not subject to the rules of ordinary arithmetic. He explains how imaginary numbers are multiplied, added, and subtracted. In it, he also looks again at the cubic problem that had perplexed Cardano—to find the solution of $x^3 = 15x + 4$ using the formula that Tartaglia had originally discovered, and that Cardano modified. This modification by Cardano was known to be the correct one to use to solve cubic problems similar to $x^3 = 15x + 4$. You will recall that when Cardano's modified formula is applied to the cubic equation, the formula reduces to $x = \sqrt[3]{2+\sqrt{-121}} - \sqrt[3]{-2+\sqrt{-121}} = \sqrt[3]{2+\sqrt{-121}} + \sqrt[3]{2-\sqrt{-121}}$.

This appears to be a formidable equation to the nonmathematical reader. But in actual fact, the equation is not difficult to understand.

Bombelli said he knew how to simplify this equation. He simply stated that $\sqrt[3]{2+\sqrt{-121}}$ is equal to $2 + i$, where $i = \sqrt{-1}$, and that $\sqrt[3]{2-\sqrt{-121}}$ is equal to $2 - i$.

We will prove Bombelli's statement here by first cubing $2 + i$. That is, we will find the value of $(2 + i)^3$. The result should equal $\sqrt{2+\sqrt{-121}}$. Then we will find the value of $(2 - i)^3$. If that equals $\sqrt{2+\sqrt{-121}}$, we know that Bombelli's statement is correct.

First we will we find the value of $(2 + i)^2$, and then we will multiply that result by $2 + i$. We find that $(2 + i)^2$ equals $4 + 4i + i^2$. Since i^2 equals -1, $4 + 4i + i^2$ equals $3 + 4i$.

Now we multiply $3 + 4i$ by $2 + i$. This equals $6 + 11i + 4i^2$. Thus $(2 + i)^3 = 6 + 11i + 4i^2$. Once again we note that $i^2 = -1$. Thus we find that $6 + 11i + 4i^2$ equals $2 + 11i$.

Thus we have proved that $(2 + i)^3$ equals $2 + 11i$.

But Bombelli's statement inferred that $(2+i)^3 = 2+\sqrt{-121}$.

Rewriting $(2 + i)^3$ as $2 + 11i$, we obtain $2+11i = 2+\sqrt{-121}$.

Therefore,

$$11i = \sqrt{121} \cdot \sqrt{-1}$$

or

$$11i = 11 \cdot \sqrt{-1}$$

or

$$i = \sqrt{-1}.$$

We have just shown that

$$(2+i)^3 = 2+\sqrt{-121}.$$

If you care to check, you will find that

$$(2-i)^3 = 2-\sqrt{-121}.$$

So we have now proved that Bombelli's solution of this cubic equation is possible, provided we accept that the square root of -1 exists. Therefore, we realize that not only does the square root of -1 turn up in the solution of a cubic equation, but its existence is essential to solving this particular cubic equation.

Therefore, the result of Cardano's formula when applied to the cubic equation $x^3 = 15x + 4$ equals $(2 + i) + (2 - i)$. If you sum the two complex numbers $2 + i$ and $2 - i$, the imaginary parts cancel each other and the real parts add to 4. Thus the solution to the cubic equation $x^3 = 15x + 4$ is 4. In other words, by making $x = 4$, the equation is correct.

Lo and behold, that is the solution to the cubic equation that perplexed Cardano!

The fact that the real solution is found using Cardano's formula created its own philosophical problem. Recall that we started with a cubic equation. The numbers in the cubic are real, everyday numbers. The solution to the cubic is 4, which is also a real, everyday number. But to get to that answer we had to encounter and manipulate so-called imaginary numbers. So the journey to the solution to this cubic equation appeared to have started with real numbers, then traveled through the domain of imaginary numbers, and finally ended up with a real number as the solution. Many mathematicians must have wondered why the laws of mathematics operate in this manner.

Of course, after Bombelli established that we could use numbers such as i, where $i^2 = -1$, you could argue that the solution, 4, is in the cubic formula all along. But you had to use complex numbers to get the real number solution out. Is this not strange?

Not really. If you think about it, *every* number consists of a real part and an imaginary part. We are not used to looking at real numbers in this manner, because we do not insert the imaginary part of the number when that imaginary part is equal to zero. So it just seems to us mere mortals that it is strange that we have to use imaginary numbers to get the solution to the cubic, when the solution is what we call a real number.

Therefore, the laws of mathematics are not behaving strangely when we solve a cubic equation. It just appears that way to our minds. (Don't forget, our minds evolved to collect fruit and kill wild animals for food on the plains of Africa about 100,000 years ago. Our minds didn't evolve to solve cubic or differential equations, so it is not surprising that the average mind finds these concepts difficult to understand.) From nature's point of view, *all* numbers have an imaginary component. When we accept this, we find that there is no great mystery to imaginary numbers. Imaginary numbers are in fact as real as any other number. They are built in to the fabric of the universe.

Incidentally, in passing I will just mention there are two other solutions to the above cubic equation. These are: $x = -2 \pm \sqrt{3}$. Thus any one of the following three values will make the equation $x^3 = 15x + 4$ correct: $x = 4$; $x = -2 + \sqrt{3}$; or $x = -2 - \sqrt{3}$.

Mathematicians today define a number, let us call it x, as being a complex number if it is of the form $a + bi$, where a and b are real numbers and i is the imaginary number, which is the square root of -1. Thus a is a real number and bi is the imaginary part. If a is equal to zero, then we have an imaginary number. If, on the other hand, b equals zero, we have a *real* number, without any imaginary part, because any number, even an imaginary one, multiplied by 0 is 0. This means that every positive number is a complex number, where bi is equal to zero. As a consequence of this, the complex numbers form a larger set than the set of real numbers. Saying the same thing another way, the set of real numbers is a subset of the complex numbers. One consequently finds that the set of imaginary numbers is also a subset of the complex numbers.

The discovery of the square root of -1 (usually written as i) was a major milestone in the history of mathematics. Since the number i is defined to be equal to the square root of -1, we find the following table gives the successive powers of i:

$$
\begin{aligned}
i^1 &= i \\
i^2 &= -1 \\
i^3 &= -i \\
i^4 &= 1 \\
i^5 &= i \\
i^6 &= i^2 = -1 \\
i^7 &= i^3 = -i \\
i^8 &= i^4 = 1
\end{aligned}
$$

You can see from this table that the successive powers of i lead to a cycle where the values obtained at successive powers (beginning with i^1) are: i, -1, $-i$, and 1. The cycle then repeats.

This means that to calculate any higher power of i, you can convert it to a lower power by taking the closest multiple of 4 that's no bigger than the exponent and subtracting this multiple from the exponent. For example, suppose you are asked to simplify i^{99}. With a little insight, this problem is easily solved by the use of a shortcut. An example will make it clear how the shortcut works: $i^{99} = i^{96} \cdot i^{3} = i^{(4 \cdot 24)} \cdot i^3 = 1 \cdot i^3 = i^3$.

Having found that $\sqrt{-1} = i$, you may wonder if some other imaginary number is required to calculate the square root of *i*. But this is not the case. The two square roots of *i* are easily found. They are

$$\sqrt{i} = \frac{1+i}{\sqrt{2}} = or\sqrt{-i} = \frac{-1+i}{\sqrt{2}}.$$

To understand how the $\sqrt{i} = \frac{1+i}{\sqrt{2}}$, consider the following: we square both sides of the equation to obtain $i = \frac{(1+i)^2}{2}$. This equals $i = \frac{2i}{2}$, which reduces to $i = i$. We do the same with $\sqrt{-i} = \frac{-1+i}{\sqrt{2}}$. Squaring both sides of the equation and simplifying, we obtain $-i = \frac{-2i}{2}$, which reduces to $-i = -i$.

The discovery of *i* means that any algebraic equation can be solved without discovering or inventing another imaginary number.

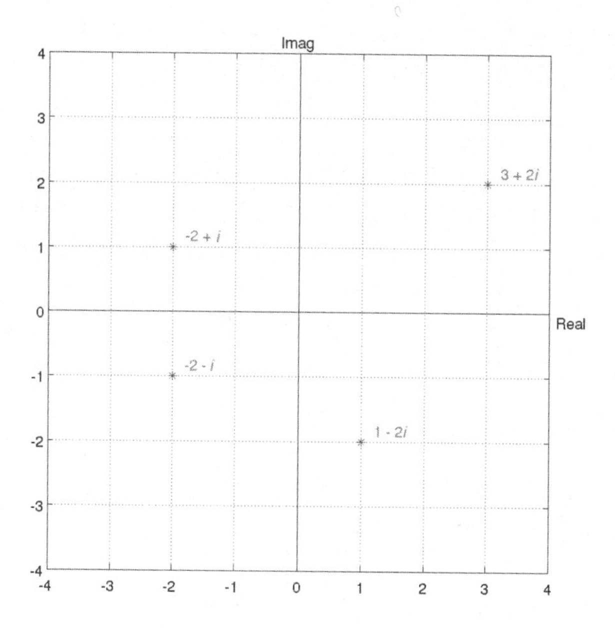

Figure 8.1. The Argand diagram.

The Argand diagram (see in figure 8.1) may be used to give a geometrical picture of the meaning of *i*.

Consider the horizontal number line (call it the *real axis*) as extending from west to east. The center point is the origin of the number line and is marked "0." At this point, let a vertical line (call it the *imaginary axis*) extend from north to south. Mark the positive numbers, 1, 2, 3, 4, 5, and so on, at equal intervals to the right of the zero. Mark the negative numbers, –1, –2, –3, –4, –5, and so on, at equal intervals to the left of the zero.

Think of the numbers $+i$, $+2i$, $+3i$, $+4i$, . . . as being at equal intervals as you move north on the northern imaginary axis, and think of the numbers $-i$, $-2i$, $-3i$,$-4i$ . . . as being at equal intervals as you move south on the southern imaginary axis. In other words, the imaginary numbers above the horizontal axis (i.e., the real axis) are positive and the imaginary numbers below the real axis are negative.

Prior to the invention (or discovery) of complex numbers, the number line extended from west to east. In other words, the number line and therefore all numbers were linear. But the discovery of complex numbers shows that the vast majority of numbers are not linear. They appear above and below the real axis. The real part of a complex number can be considered to be a displacement along the real axis (the x or horizontal axis), and the complex part of the number a displacement along the imaginary axis (the y or vertical axis). (Every number on the positive axis is on the complex plane.)

It is best to consider the multiplication process involving complex numbers in the following manner. Every time we multiply a number on the *complex plane* by +1, we are performing an operation on the number that is being multiplied so that that number is rotated 360°across the complex plane. Every time we multiply by –1, we are performing an operation on the number that is being multiplied so that that number is rotated 180° on the number line.

Let's see what happens when we multiply a number on the complex plane by i. It merely rotates the number that is being multiplied by 90° counterclockwise across the complex plane. (See figure 8.2.)

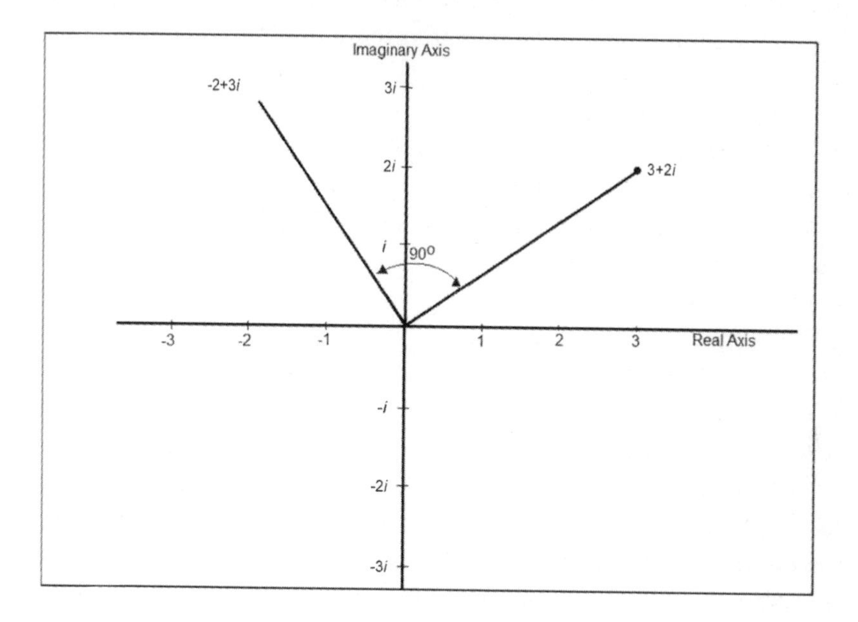

Figure 8.2

Figure 8.2 shows the location of the complex number $3 + 2i$. If you multiply $3 + 2i$ by i, keeping in mind that i^2 equals -1, you obtain $-2 + 3i$. In other words, the location of the complex number $3 + 2i$ has been rotated 90° across the complex plane by the fact that it was multiplied by i. Since most of us make at least one 90° rotation as we go on our travels every day, we see that the operation of multiplying a number by i involves a similar rotation.

One of the curious properties of i is that it can be easily shown to be related to primitive Pythagorean triples. (We discussed these in chapter 3.) You may recall that these are integers that satisfy the Pythagorean theorem: $a^2 + b^2 = c^2$. When the numbers a, b, and c are reduced so that there is no common factor between them (meaning that they are not divisible by the same number), the three numbers form a *primitive* Pythagorean triple. The first five primitive Pythagorean triples are:

$$3^2 + 4^2 = 5^2$$
$$5^2 + 12^2 = 13^2$$
$$7^2 + 24^2 = 25^2$$

$$9^2 + 40^2 = 41^2$$
$$11^2 + 60^2 = 61^2$$

To see how these triples are related to the complex numbers, consider the following pattern involving complex numbers:

$$(1 + 2i)^2 = -3 + 4i \quad > \quad 3^2 + 4^2 = 5^2$$
$$(2 + 3i)^2 -5 + 12i \quad > \quad 5^2 + 12^2 = 13^2$$
$$(3 + 4i)^2 = -7 + 24i \quad > \quad 7^2 + 24^2 = 25^2$$
$$(4 + 5i)^2 = -9 + 40i \quad > \quad 9^2 + 40^2 = 5^2$$
$$(5 + 6i)^2 = -11 + 60i \quad > \quad 11^2 + 60^2 = 61^2$$
$$\ldots$$

We encountered earlier in this book the irrational number phi (Φ), which equals 1.618033988 . . . . (The reader may recall that $\Phi^2 - 1$ equals Φ.) Who would have thought, for instance, that Φ is intimately connected to i? But it is, as is shown in the following simple equation:

$$\frac{\Phi + i}{\sqrt{\Phi}} = (\sqrt{1+2i}).$$

To prove that this identity is true, square both sides of the equation. This gives

$$\frac{\Phi^2 + 2i\Phi + i^2}{\Phi} = 1 + 2i.$$

Since $i^2 = -1$, this allows us to write

$$\frac{\Phi^2 + 2i\Phi - 1}{\Phi} = 1 + 2i.$$

Since $\Phi^2 = \Phi + 1$, this in turn equals

$$\frac{\Phi + 1 + 2i\Phi - 1}{\Phi} = 1 + 2i.$$

This equals

$$\frac{\Phi + 2i\Phi}{\Phi} = 1 + 2i,$$

which reduces to

$$1 + 2i = 1 + 2i.$$

Therefore, the original equation $\frac{\Phi + i}{\sqrt{\Phi}} = (\sqrt{1+2i})$ is true.

Complex numbers crop up as solutions to many problems in mathematics. For example, consider the following geometrical problem. Begin with a quadrilateral. Then, construct a square on each of the four sides of the quadrilateral. Draw lines from the center of each square to the center of the opposite square. (See figure 8.3.)

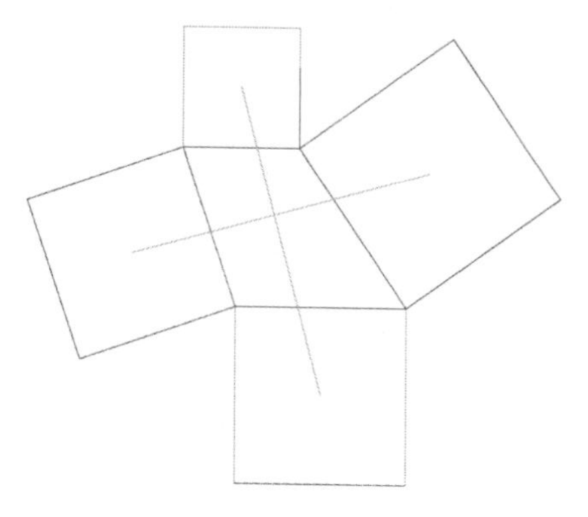

Figure 8.3

The problem is to prove that these lines are perpendicular and are of equal length. In his brilliant book *Visual Complex Analysis*, Tristan Needham gives a beautiful solution to this problem using complex numbers.[3] He then goes on to outline an alternative proof of the theorem without using complex numbers. This alternative proof is much longer than the proof based on complex numbers. So, in some way, it can be said that the more natural, simple proof is the one that makes use of complex numbers.

Another problem that is solved with the use of complex numbers is the following. Suppose we construct four points equally spaced around a unit circle. We then draw chords *AD*, *BD*, and *CD* from one of the points to each of the other three points. (See figure 8.4.) What is the product of the length of those three chords?

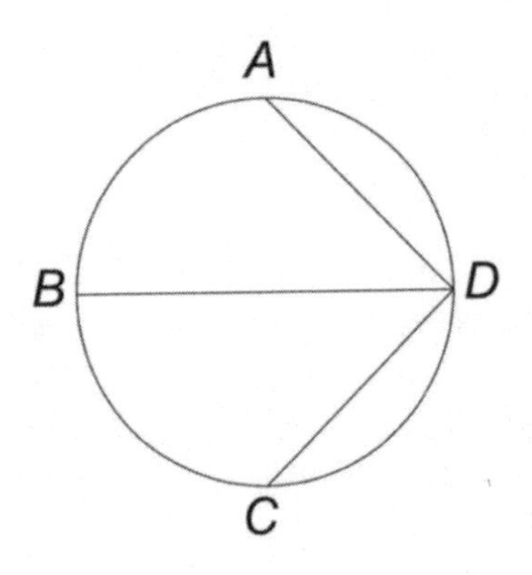

Figure 8.4

The surprising result is that the product of the three chords *AD*, *BD*, and *CD* equals 4, which is the number of points around the circle that we started with. In other words, starting with n points equally spaced around the circumference of a unit circle, the product of the $n - 1$ chords is n. Is this always the case, regardless of the number of points?

The answer is yes. In general, divide the circumference of a circle whose radius is 1 unit into n equal arcs, or points. Draw chords from one of the points to the other $n - 1$ points. Let $c_1, c_2, c_3, \ldots$ represent the chords from one of the points to the other $n - 1$ points. It can be mathematically proved that the product of the $n - 1$ chord lengths equals n. The proof of this involves complex numbers. I do not know if it can be proved without using complex numbers, but if it can, I would surmise that it is a much longer and probably more onerous proof.

A famous result in number theory states that if m and n are both integers where both m and n equal the sum of two squares, then mn can also be expressed as the sum of two squares. For example, 5 and 13 can both be expressed as the sum of two squares: $5 = 1^2 + 2^2$ and $13 = 2^2 + 3^2$. The product of 5 and 13 equals 65, and $65 = 1^2 + 8^2$. This result can also be proved with the use of complex numbers.

One also encounters i in the study of matter and energy. For instance, the fundamental equation of physics for describing mechanical behavior is known as the *Schrödinger equation*. Incredibly, that equation includes the number i. It seems that nature, at its most basic level, cannot be mathematically described in an elegant manner without involving the square root of –1.

Astrophysicist Freeman Dyson (1923–) famously commented on this fact. He said that it was one of the most profound jokes of nature that the square root of –1 was the identity the physicist Erwin Schrödinger (1887–1961) put into his wave equation when he invented wave mechanics in 1926. Dyson declared that he believed that the mathematicians of the nineteenth century viewed the square root of –1 as an artificial construction, invented by human mathematicians as a useful device that helped solve equations. Dyson goes on to say that it never entered the heads of these mathematicians that the imaginary numbers that they thought they had *invented* were in fact the ground in which atoms move. Nature, Dyson famously said, got there first.[4]

Many of the modern marvels of technology are based on mathematical equations that include the number i. Computers, CDs, cell phones, digital cameras, and numerous other devices would not exist, were it not for the fact that the square root of –1 is accepted and used in our mathematics.

Here is another astonishing property of i: i to the i power equals 0.20787957. . . .

That is, $i^i = 0.20787957 \ldots$.

In fact, it can be proved that i^i has an infinite number of possible values. The value of i^i being equal to 0.207879 . . . is known as the *principal* value of i^i.

This is a truly amazing property of i. Here we have a so-called imaginary number (a number whose existence many still refuse to accept to this day) raised to the power of an imaginary number, and the result is a number slightly greater than 0.2. What an extraordinary result! Who would have thought that a number—invented or discovered, depending on your point of view about mathematical objects—that equals the square root of –1 and is then raised to its own power equals a real number that is slightly more than 0.2?

This property of i has been looked upon with awe and fascination for decades by mathematicians, philosophers, and mystics. Why should an imaginary number raised to the power of an imaginary number equal a positive number? It seems to go against common sense. It is probable that some mystics have looked at this equation and wondered if it is a hint from nature that she can bring about a real universe from nothing!

The equation becomes even more astonishing when you learn that $i^i = e^{(-\pi/2)}$. In other words, i raised to the power of i equals the number known as e raised to the power of $-\pi/2$. Here the number represented by the lower case e is a famous number in mathematics. The number e equals 2.718281828 . . . . Its decimal expansion goes on forever. The other number that appears in the equation is π. (Like π, the number e is ubiquitous in mathematics. We will discuss *e and* π later in this book.)

The above identity, i^i, can also be expressed as follows:

$$i^i = \left(\frac{1}{\sqrt{e}}\right)^{\pi}.$$

The above equation shows a surprising connection between i, e, and π. The *logarithm* of i can also be calculated:

$$\ln(i) = i \cdot \frac{\pi}{2}.$$

Who would have thought that the natural logarithm of i would involve the number 3.14159265 . . . ? Who could have foreseen the link between natural logarithms, π, and the number i?

I sent a draft of this chapter to Dr. Cong, and he sent me the following reply.

Hi, Owen,

Thank you for sending me your discussion of the imaginary number i. As you quite correctly point out, there is nothing imaginary about i. It is as real as any other number. Bombelli's book appeared in 1572. His was the first book in history to illustrate how negative numbers can be used to obtain a solution to a cubic equation in real numbers.

Incidentally, your readers should be made aware that electrical engineers use the square root of -1 extensively in their work, but they denote it as j, rather than i, because i is usually used to represent current.

Many curious properties of i have been discovered. One simple example is that when i is multiplied by -1, it equals its own reciprocal. In other words,

$$i \cdot -1 = i \cdot i^2 = i^3 = i^4/i = 1/i.$$

One curious feature of complex numbers that you did not mention in your chapter is that when two or more complex numbers are multiplied together, the angles between the resulting complex numbers and the original complex numbers are summed. This means that if you square a complex number, the resulting complex number has twice the angle of the original number.

Let's take an example of this. Suppose you square $3 + 4i$. The result is $-7 + 24i$. Let us say that $3 + 4i$ equals A, and let us say that $-7 + 24i$ equals B. The length of the diagonal going from zero to the complex number $3 + 4i$ is, by the Pythagorean theorem, equal to $\sqrt{3^2+4^2}$, which equals 5. The length of the diagonal going from 0 to the complex number $-7 + 24i$ is $\sqrt{7^2+24^2}$, which equals 25.

The angle of A is the arctan of $\frac{4}{3}$. This equals 53.13010235°. (In other words, the diagonal of a rectangle that is 3 units long and 4 units wide makes an angle of 53.13010235°with the base.) The angle of B equals 180 minus the arctan of $\frac{24}{7}$. This equals 180 – 73.73979529°, or 106.2602047°. You can see that 106.2602047° is just twice 53.13010235°.

The above information can be very helpful in solving certain types of geometrical problems. For example, consider the following little puzzle. What is the sum of the angles of A, B, and C in the below figure [figure 8.5]?

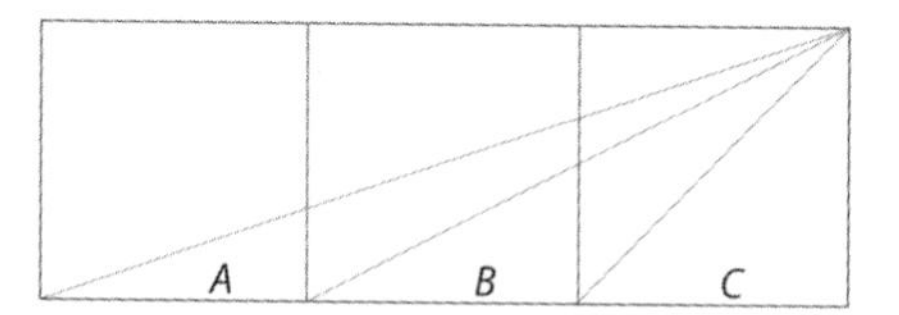

Figure 8.5

It can be solved in a variety of ways. However, by using complex numbers it can be solved very easily. Ask your readers if they can solve this little teaser using complex numbers. Perhaps the tip I give about the angles being summed when two or more complex numbers are multiplied together will be helpful to them.

Before I go, I'll give an example of a little number play with the number 1,572, which was the year Bombelli's *Algebra* was published.

Consider the number 1,572. Partition this into the two parts 15 and 72. The 72nd prime number is 359. The product of the digits in the number 359 is 135. Multiply that number by its end digit, and you obtain 675. Multiply that number by its end digit, and you obtain 3,375, which is the *cube* of 15.

The sum of the digits of 1,572 is 15, and the square of 15 is 225. Add 225 to its reversal, and you obtain 747. Move the 7 at the front of the number to the end to obtain the number 477. The sum of the *cubes* of the digits of 1,572 is 477.

Your readers might like to learn that complex numbers can be used to solve other mathematical puzzles that turn up in recreational mathematics. A famous example of this is the puzzle of the buried treasure. This puzzle has appeared so many times in books on recreational mathematics that I will not repeat the problem here. Interested readers can learn more about it from the following sources:

- George Gamow's *One Two Three . . . Infinity* (New York: Viking, 1947; rpr. New York: Dover, 1988.) Here, George Gamow presents the puzzle and solution.
- Paul J. Nahin's *An Imaginary Tale: The Story of* $\sqrt{-1}$ (Princeton, NJ: Princeton University Press, 1998). In this piece, Nahin provides the same puzzle but a slightly different method of solution.
- Martin Gardner's "Imaginary Numbers," chap. 17 in *Fractal Music, Hypercards and More Mathematical Recreations from Scientific American Magazine* (New York: W. H. Freeman, 1992).

I better sign off now.

Talk to you soon.

Best regards,

Dr. Cong

SOLUTION

(1) The puzzle about the sum of the three angles shown in figure 8.5 is solved using complex numbers as follows:

Angle A is formed by a line going 3 units east and 1 unit north. Therefore, the point that corresponds to this angle can be represented on the complex plane by the complex number $3 + i$. Similarly, angle B is formed by a line going 2 units east and 1 unit north. The point that corresponds to this angle can be represented on the complex plane as $2 + i$. Angle C is formed by a line going 1 unit east and 1 unit north, which can be represented on the complex plane as $1 + i$.

We know from Dr. Cong's hint that when two or more complex numbers are multiplied together, the angles between the resulting complex numbers and the original complex numbers are summed.

We thus multiply $3 + i$ by $2 + i$, and then that result by $1 + i$. The result is $10i$. This point, $10i$, lies on the vertical (y) axis on the complex plane and therefore makes a 90° angle with the origin (the intersection of the x axis and the y axis). Now the point $1 + i$ on the complex plane makes a 45° angle with the origin. Dr. Cong's hint tells us that the other two angles summed must equal 90° minus 45°. Therefore, the sum of the other two angles also equals 45°.

Thus we are left with the surprising solution that angle A + angle B equals angle C. Since angle C equals 45°, angle A + angle B equals 45°, and the sum of the three angles, A, B, and C, equals 90°.

CHAPTER 9

THE SQUARE ROOT OF 2

Suppose we construct on a sheet of blank paper a square whose length is an exact integer. The question can now be asked, What is the *exact* length of the diagonal of such a square?

The question seems sensible to a nonmathematician, and there appears to be no reason why the exact length cannot be computed. The square is drawn on the page and the exact length of each side of the square is one unit, let us say one foot. So the length of the diagonal is . . . yes, it is calculated by using the Pythagorean theorem that applies to all right triangles The sum of the squares of the two legs is equal to the square on the hypotenuse. Therefore, the square on the hypotenuse has an area of 2 square units. The length of that square is equal to the square root of 2, which is about 1.414 . . . feet long. That is one foot, 4 inches, and about 9/10 of an inch long.

Suppose, however, we want the *exact* length? Let's see what can be done. Perhaps we can measure the diagonal a little more accurately. Suppose we have a strong magnifying glass and magnify the square to, say, one hundred times its length. Each side of the square is now one hundred feet. The length of the diagonal of this square is one hundred times the square root of 2. This equals 141.421 . . . . Thus the diagonal of this enlarged square works out to be 141 feet, 5 inches, and about 5/100 of an inch long. Ah, yes, we have now calculated that the length of the diagonal is about 141.421 feet in length. But the mathematician will tell us that that is not the *exact* length of the diagonal. Why? Well, by the Pythagorean theorem, the square on the hypotenuse of a triangle whose two sides are 100 feet in length has an area of 20,000 square feet. Therefore, the length of the diagonal must be some number that, multiplied by itself, gives 20,000. The square root of 20,000 is 141.421356 . . . . The value we obtained as the length of the diagonal is 141.421. The value we obtained is close to the true value, but it is not the *exact* length of the diagonal. So the question reduces to this: What is the exact length of the diagonal?

In this day and age, when we have split the atom, sent astronauts to the moon, and have developed extremely powerful computers, we may well believe that it is not beyond human ingenuity to have the technology to measure the *exact* length of the diagonal of a square.

But the mathematician smiles and shakes her head. She then proceeds to tell us gently that we cannot ever measure the *exact* length of the diagonal of a square if the square side is equal to an integer or a rational number. (A rational number is one that can be expressed as a/b, where a and b are both integers. A square having rational sides equal to a/b can easily be enlarged to a side having integer sides. So if we could measure exactly the diagonal of a square with rational sides, we would be able to measure the exact length of a diagonal in any square.)

The reason we cannot measure the diagonal of a square *exactly* is not because we do not have the technology to do so; it is because the *exact* length of a diagonal of a square which has integer sides (or rational sides) *does not exist in relation to the length of the sides of the square*! Let me say that again. In a square that has sides equal *exactly* to an integer or a rational number, the *exact* length of the diagonal of the square *in relation to the length of a side* does not exist!

The mathematician describes this situation by saying that the length of the diagonal of a square is *incommensurable* with the sides if the length of each of the sides is an integer or a rational number. So if the side of the square is one foot in length, the mathematician will say that the length of the diagonal is 1.4142135 . . . feet. The number 1.4142135 . . . is usually described by the mathematician as the square root of 2. Sometimes it is just called "root of 2." The symbol the mathematician uses for the square root is $\sqrt{\ }$. That symbol comes from the Latin letter *R*, which represents the word *radix*, which means "root." Thus the square root of 2 is usually written as $\sqrt{2}$.

The reader may recall that the Pythagorean theorem tells us that in any right triangle the area of the sum of the squares on each of the two legs is equal to the square of the hypotenuse. So let's say the length of each leg in a right triangle is one unit. We are not told the length of the long side (the hypotenuse) of the right triangle. Let's just say at this moment that it is z feet long. According to the Pythagorean theorem, $1^2 + 1^2 = z^2$. Therefore, the area of the square on the hypotenuse is 2 square feet. Therefore, the length of the hypotenuse is a number that when multiplied by itself *exactly* equals 2. This number is therefore described as $\sqrt{2}$.

It can easily be proved that $\sqrt{2}$ cannot be expressed as a fraction. (If it could, there would no problem determining the exact length of the diagonal of a square.) We know for instance that 7/5 is approximately $\sqrt{2}$. If you square 7/5, you obtain 49/25. So although $7^2/5^2$ is relatively close to the value of 2, it is not *exactly* 2. Therefore, the fraction 7/5 is not exactly the $\sqrt{2}$. The fraction 99/70 is closer *still* to $\sqrt{2}$, but it is not *exactly* it either. We can find fractions that will give closer and closer values to $\sqrt{2}$ but will not give the *exact* value, because it can be mathematically proved that no such fraction exists.

Here is an ancient, beautiful proof that $\sqrt{2}$ is irrational. If it was rational, it could be expressed as a/b, where a and b are both integers. Let's assume that such a fraction exists, but we have reduced the fraction so it has no common factors. Therefore we can write:

$$\frac{a}{b} = \sqrt{2}$$

or

$$\frac{a^2}{b^2} = 2.$$

Consequently, $a^2 = 2b^2$. The right-hand side of this equation is divisible by 2. That means that the left-hand side is also divisible by 2. So the left-hand side must be even. Now a^2 can only be even if a is even. If a is even, then a^2 is divisible by 4. Thus the right-hand side of the equation must be divisible by 4. That means b^2 is even, which in turn means that b is even. So both a and b are divisible by 2. But we initially assumed that a and b did not have a common factor. We now have a contradiction. Therefore, our original assumption that a fraction equaling $\sqrt{2}$ exists must be false.

The square root of 2 can only be expressed accurately as the limit of an infinite series, or as the limit of an infinite continued fraction. In years gone by, the square root of any integer would be calculated by a tedious mathematical procedure. Today, however, computer programmers can program the relevant mathematical formulae into computers to obtain the square root of any number to as many decimal digits as may be required. We know, for example, that the square root of 2 to the first seven decimal places is 1.4142135. This may be rewritten as follows:

$$\sqrt{2} = 1 + \frac{4}{10} + \frac{1}{100} + \frac{4}{1000} + \frac{2}{10000} + \frac{1}{100000} + \frac{3}{1000000} + \frac{5}{10000000} + \ldots$$

It can also be expressed as the following infinite continued fraction:

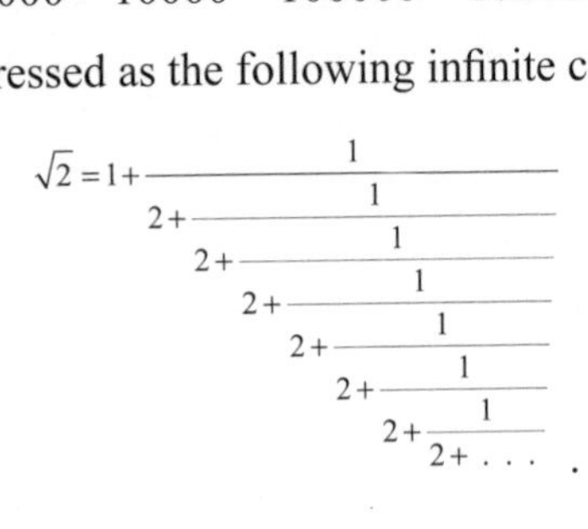

$$\sqrt{2} = 1 + \cfrac{1}{2 + \cfrac{1}{2 + \cfrac{1}{2 + \cfrac{1}{2 + \cfrac{1}{2 + \cfrac{1}{2 + \cfrac{1}{2 + \ldots}}}}}}}.$$

There are, of course, an infinite number of irrational numbers, but $\sqrt{2}$ is the first to have been discovered. These numbers are called ir*ratio*nal because they cannot be expressed as a *ratio* between two integers. That such numbers could exist came as a great shock to mathematicians at the time they were discovered. But now mathematicians constantly use irrationals and their existence is no longer in any doubt.

Because $\sqrt{2}$ is an irrational number, its decimal expansion goes on forever. Think about that for a moment. Here is a number, 1.4142135 . . . , whose decimal digits never ever end. And the decimal digits never repeat. (If they did, the fraction would not be irrational.) The fraction for 1/3, for example, when expressed as a decimal, equals 0.33333333 . . . . The number of 3s goes on forever. But the decimal expansion for $\sqrt{2}$ does not appear to have a pattern. For example, if we were given the following ten digits, 7187537694, there is no apparent way that we could tell that these digits are the forty-first to the fiftieth digits of $\sqrt{2}$.

If we were to write down the decimal digits of $\sqrt{2}$, we could do so forever, without ever getting to its last decimal digit. Why? Because $\sqrt{2}$ does not have a last decimal digit! Yet, incredibly, we know that when this number—whose decimal digits continue forever—is multiplied by itself, the answer is *exactly* 2. Who would have thought that such a thing is possible?

Here are the first thirty decimals of $\sqrt{2}$:

1.414213562373095048801688724209.

There are numerous curiosities involving $\sqrt{2}$ that are unknown to many mathematics teachers. Most of these curiosities are easily understood, but for some reason they have never attracted the attention they deserve. For example, students of mathematics are often surprised when they first learn that 1 divided by $\sqrt{50}$ equals 0.141421356237 . . . . The decimal digits of this number are identical to all the digits of $\sqrt{2}$. This curiosity arises because the following equation holds:

$$100 = 2 \cdot 50.$$

Therefore,

$$10 = \left(\sqrt{2}\right)\left(\sqrt{50}\right).$$

Divide both sides of the equation by $\sqrt{50}$:

$$\frac{10}{\sqrt{50}} = \sqrt{2} \qquad \sqrt{50}$$

Divide both sides of the equation by 10:

$$\frac{1}{\sqrt{50}} = \frac{\sqrt{2}}{10}$$

This result equals

$$\frac{1}{\sqrt{50}} = \frac{\sqrt{2}}{10} = \frac{1.4142135\ldots}{10} = 0.14142135\ldots$$

or

$$\frac{1}{\sqrt{50}} = \frac{1.4142135\ldots}{10} = 0.14142135\ldots.$$

This is the curiosity we gave above.

The square root of 2 has many other interesting properties. Here are just two of these:

$$\frac{1}{\sqrt{2}} = \frac{1}{2} \cdot \sqrt{2} \text{ and } \left(\frac{1}{\sqrt{2}}\right)^2 = \frac{1}{2}.$$

The decimal part of $\sqrt{2}$ has been calculated to 200 million decimal places by computer programmers around the world. The reasons programmers perform these calculations differ from individual to individual. Of course they are all well aware that the decimal expansion is never ending. So why do they do it?

Some do so because they like to search for any unusual pattern in the endless line of digits. Others do so to see if $\sqrt{2}$ is a normal number. A normal number is one in which the endless decimal digits are distributed evenly. For instance, the number 1 should appear about one tenth of the time in the decimal expansion of $\sqrt{2}$. So should any other single digit. A double-digit number, such as 23, should appear about one-hundredth of the time in the decimal expansion. A triple-digit number, such as 345, should appear about one thousandth of the time in the decimal expansion. And so on. From checks performed on the decimal part of the root of 2, it does seem to be a normal number, but this has not yet been proven. If it is ever proven to be a normal number, mathematicians believe that the mathematical techniques required to establish the proof will probably be very advanced. On the other hand, there is always that small chance that there just may be a relatively easy method—

unknown to mathematicians at this moment in time—that may yield an answer to this question.

Why does it matter whether or not $\sqrt{2}$ is normal? Well, mathematicians hope that the answer one way or the other will tell us more about the nature of irrational numbers. For all we know, there may be some deep mathematical reason why these numbers are normal, or not normal, as the case may be. (If there is such a deep reason and we discover it, it will expand humanity's knowledge of number theory and may also lead to other deep results in mathematics. That in turn will hopefully eventually bring benefits to the human family.) Of course, there is also a human element to this quest. If it is possible for human beings to find out if $\sqrt{2}$ is or is not a normal number, then some human beings will do their utmost to find a solution to the problem.

For all we know, there may be a deep reason why $\sqrt{2}$ is a normal number (or is not a normal number), which is related to the fact that there is no known shortcut to factoring large semiprime numbers. The security of Internet transactions depends on the fact that no such shortcut to factoring exists. (See chapter 2.) If, therefore, we gain some knowledge about the normality or otherwise of $\sqrt{2}$ and other irrational numbers, it can only help in our attempts to understand other mathematical conundrums.

The square root of 2 turns up in many beautiful formulae in mathematics. For example, the following infinite series equals $\sqrt{2}$:

$$\sqrt{2} = \frac{2 \cdot 2}{1 \cdot 3} \cdot \frac{6 \cdot 6}{5 \cdot 7} \cdot \frac{10 \cdot 10}{9 \cdot 11} \cdot \frac{14 \cdot 14}{13 \cdot 15} \cdot \frac{18 \cdot 18}{17 \cdot 19} \cdot \frac{22 \cdot 22}{21 \cdot 23} \cdot \ldots$$

The denominators in the above equation are all of the odd integers in order, beginning with 1.

The following equation also equals $\sqrt{2}$:

$$\sqrt{2} = 1 + \frac{1}{2} - \frac{1}{2 \cdot 4} + \frac{1 \cdot 3}{2 \cdot 4 \cdot 6} - \frac{1 \cdot 3 \cdot 5}{2 \cdot 4 \cdot 6 \cdot 8} \cdot$$

Who would have thought that the length of the diagonal of a square with sides equal to one unit could be part of such beautiful equations?

Many high-school students studying trigonometry are often surprised when they first learn of the following equation:

$$\sin 45° = \frac{\sqrt{2}}{2} = \cos 45°.$$

The sine of an angle is the relation between the long side of a triangle and the hypotenuse. In a 45-degree angle, both legs of the triangle are equal in length. If the hypotenuse opposite a 45-degree angle is 1 unit in length, each of the legs is 0.70710678 . . . in length. 0.70710678 . . . equals $\sqrt{2}$ divided by 2.

The square root of 2 also appears in this unusual-looking equation:

$$\left(\sqrt{2}+1\right)\left(\sqrt{2}-1\right)=1.$$

As can be expected, $\sqrt{2}$ turns up in various figures in plane geometry. One classic case is in the octagon, an eight-sided figure with eight equal angles that add up to 1080 degrees (see figure 9.1). Incidentally, the sum of the interior angles of a polygon with n sides is 180 times $(n-2)$. Thus the sum of the interior angles of an octagon is 180 times 6, which equals 1,080.

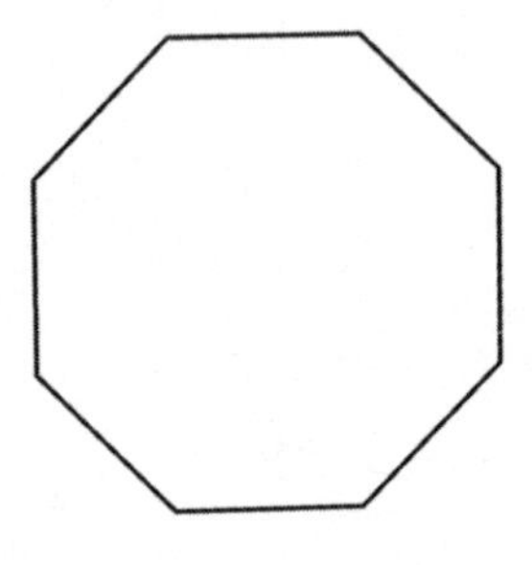

Figure 9.1

If a equals the length of one of the eight sides of an octagon, the formula to calculate the octagon's area is $2a^2(1+\sqrt{2})$. Therefore, if a equals 1, the area of the octagon is 4.8284 . . . . It is possible to draw diagonals of three different lengths (call these lengths long, medium, and short) inside the octagon. If the length of the side of the octagon is 1, the lengths of the three diagonals will equal the following:

long diagonal:

$$a \cdot \sqrt{4+2 \cdot \sqrt{2}}$$

medium diagonal:

$$a \cdot \left(1+\sqrt{2}\right)$$

short diagonal:

$$a \cdot \sqrt{4+2 \cdot \sqrt{2}}$$

Interested readers may enjoy working out the values of these identities themselves.

The square root of 2 appears in the solutions of problems that arise in the commercial world. For instance, the international standard of paper sizes used in Europe today for office work such as photocopying or printing is specified by the International Organization for Standardization (ISO). The most common sizes available (outside of the United States of America) are the "A" series.

The series starts off with A0, A1, A2, A3, A4, A5, up to and including A10. The size of the A0 paper is 841 mm in width and 1,189 mm in length. The size of the A1 is 594 mm in width by 841 mm in length; A2 is 420 mm by 594 mm; and A3 is 297 mm by 420 mm.

The next size is probably the most widely used sheet of paper in Europe and probably in the world. It is the A4 size, which measures 210 mm in width by 297 mm in length. Readers may well ask why these dimensions are chosen for these particular paper sizes.

A quick check of the dimensions will show that the length of each paper sheet divided by the width is about 1.4. The dimensions of the A0 sheet are 1,189 and 841. Divide 1,189 by 841, and you obtain 1.4137931 . . . . The dimensions of the A1 paper sheet are 841 by 594. Divide 841 by 594, and you obtain 1.4158249 . . . . And so on with the other "A" sheets.

You will notice that these ratios are fractions that are very close approximations to $\sqrt{2}$. This is no coincidence. These dimensions are deliberately chosen to equal the root of 2. Why? Because if you cut or fold a sheet of "A" paper in half widthwise, the resulting halves have the same ratio of length to width as the original piece. For example, if you cut the A4 piece in half widthwise, the dimensions of the two resulting pieces will be 148.5 mm in length and 210 mm in width. If you divide 210 by 148.5, the result is 1.4141414 . . . , which is very close to the root of 2. This property of the "A" series only works because the ratio of the length of the piece of paper to its width is very close to $\sqrt{2}$. If it were any other ratio, this property of retaining the ratio when the paper is cut or folded in half would not hold.

Incidentally, as a consequence of using this unique ratio, the area of each sheet of paper in the "A" series is half of the next larger size. For example, the area of the A4 sheet is 210 by 297, or 62,370 square millimeters. The area of the next size, A3, is 297 by 420, or 124,740 square millimeters. And so on. The ISO "B" series and "C" series have different sizes to the "A" series, but their dimensions are also in the $\sqrt{2}$ to 1 ratio. Thus the B4 sheet, for example, measures 353 by 250; and the C5 sheet 229 by 162.

The square root of 2 is intimately connected to numbers that are both triangular and square. (We will be discussing triangular numbers and square numbers in more detail in later chapters.) The series of triangular numbers is 1, 3, 6, 10, 15, 21, 28, 36, 45, 55, 66 . . . . The series of square numbers is 1, 4, 9, 16, 25, 36, 49, 64, 81, 100, 121 . . . . The series of numbers that are both triangular and square is 1; 36; 1,225; 41,616; 1,413,721; 48,024,900; 1,631,432,881 . . . . Consider the table shown in figure 9.2.

*n*th triangular and square number	*n*th square	*n*th triangular	ratio of *n*th square to *n*th triangular
1	1	1	1
36	6	8	1.333333333 . . .
1225	35	49	1.4
41616	204	288	1.411764706 . . .
1413721	1189	1681	1.413793103 . . .
48024900	6930	9800	1.414141414 . . .
1631432881	40391	57121	1.414201183

Figure 9.2

We can see that the ratio of the *n*th square to the *n*th triangular number approaches the value of $\sqrt{2}$. In addition, the ratio of successive numbers that are both square and triangular approaches 33.97056275 . . . . This number equals $(1 + \sqrt{2})^4$.

For example, 1 divided by 36 equals 0.0277777 . . . . When 1,225 is divided by 36, the result is 34.0277777 . . . . When 41,616 is divided by 1,225, the answer is 33.9722448 . . . . When 1,413,721 is divided by 41,616 the result is 33.9706122 . . . . When 48,024,900 is divided by 1,413,721, the answer is 33.9705642 . . . . When 1,631,432,881 is divided by 48,024,900, the result is 33.9705627 . . . . And so on, with each successive division giving a result that is closer and closer to $(1 + \sqrt{2})^4$.

The remarkable formula for generating numbers for the *n*th number that is both square and triangular also contains $\sqrt{2}$. It was discovered by the mathematical genius Leonhard Euler in 1778. The formula is

$$\left(\frac{\left(3+2\sqrt{2}\right)^n - \left(3-2\sqrt{2}\right)^n}{4\sqrt{2}} \right)^2.$$

If you substitute 1, 2, 3, 4 . . . for n in the formula, you obtain: 1; 36; 1,225; 41,616; 1,413,721 . . . . These are the numbers that are both square and triangular.

One of the most interesting aspects of $\sqrt{2}$ is that it crops up in formulae describing natural phenomena, too. For example, consider the speed of gravity on the surface of the earth. If we were to drop a stone (starting at rest) from some height, it will drop 16.086992 . . . feet in the first second. How long will it take to drop just twice that height, that is, 32.173981 . . . . feet? The answer is $1 \cdot \sqrt{2}$ seconds, or 1.4142135 . . . , seconds.

We know that a stone starting at rest will fall 402.174802 . . . feet in five seconds. How long will it take to fall twice that distance, that is, 804.349605 . . . feet? The answer is $5 \cdot \sqrt{2}$, or 7.071067811 . . . seconds.

Suppose a stone falling from rest falls 65.6413625 . . . feet. When the stone hits the ground, it will be traveling at 64.9914448 feet per second. It will take 2.0199999 seconds to hit the ground. If the distance that the stone falls from rest is doubled, the new distance equals 131.282725 . . . feet. The stone will now hit the ground at a speed of $64.9914448 \cdot \sqrt{2}$, or 91.911787 . . . , feet per second. It will take $2.0199999 \cdot \sqrt{2}$, or 2.8567113 . . . , seconds for the stone to hit the ground.

The square root of 2 also turns up in formulae describing the motion of the planets as they orbit the sun. Mercury, the closest planet to the sun, travels at 29.8 miles per second with respect to the sun as it orbits that star. The second closest planet to the sun is Venus. Its speed around the sun is at the rate of 21.7 miles per second. The third planet form the sun is Earth. It orbits our nearest star at the rate of 18.5 miles per second.

So where does $\sqrt{2}$ come into things?

Well, all these planets orbit the sun because the sun has a gravitational pull on each of them.

For any of these planets to escape the sun's gravitational pull, they would have to increase their orbital speed. How much would that increase be? By $\sqrt{2}$ times their present speed!

So, for example, to escape the sun's gravitational pull, Mercury, whose orbital speed is 29.8 . . . miles per second, would have to increase by $29.8 \cdot \sqrt{2}$, which equals 42.1 . . . miles per second. Venus would have to increase its orbital speed from 21.7 . . . miles per second at present to $21.7 \cdot \sqrt{2}$, or 30.7 . . . , miles per second. The planet we live on, Earth, would have to increase its orbital speed from 18.5 . . . miles per second to $18.5 \cdot \sqrt{2}$, or 26.2 . . . , miles per second.

The square root of 2 does not confine itself to the orbits of only planets. It also crops up in the orbits of artificial satellites orbiting Earth or indeed any other planet, or stars orbiting other stars, and so on. Let's take an example. Suppose we here on Earth decide to launch a satellite that will orbit Earth in a circular orbit at a relatively low level. Such satellites are known as *Low Earth Orbit* (or LEO) satellites. Let's assume the altitude of the satellite is 150 miles above Earth. To travel in such an orbit, the speed of the satellite will have to be 17,360 miles per hour, or 4.8222 . . . miles per second. In this example, the satellite will orbit Earth in about 89 minutes and 20 seconds. Such a satellite is obviously captured by Earth's gravitational field. If the speed of that satellite is increased by $\sqrt{2}$ times its speed ($17{,}360 \cdot \sqrt{2}$) to 24,550.7474 . . . miles per hour, it would escape Earth's gravitational pull. That equals 6.8196 . . . miles per second.

To escape the earth's gravitational pull, or that of a planet or a moon, and leave it without further propulsion, a spacecraft needs to be traveling at a specific minimum speed. This speed is called the *escape velocity*. At the surface of the earth, the escape velocity is 6.98 miles per second. We saw that for a satellite 150 miles above Earth's surface the escape velocity is 6.8196 miles per second. This is because the earth's gravitational pull is weaker 150 miles above the surface of the planet than it is at the earth's surface.

Earth orbits the sun once every 365.25 days. Suppose a planet is discovered to exist that is twice as far from the sun as the earth is. How long would it take that planet to orbit the sun? The result is given by the third law of the three Kepler's laws of planetary motion.

In this example, we are told that a planet is discovered that is twice as far from the sun as Earth is. If Earth's distance from the sun is 1, that other planet's distance from the sun is 2. Cube 2. The result is 8. Then get the square root of 8. This equals 2.8284271 . . . . This number is exactly twice the square root of 2. It would take $2 \cdot \sqrt{2}$, or 2.8284 . . . , Earth years for this new planet to go around the sun just once. That new planet would travel at 1 divided by $\sqrt{2}$ times the orbital speed of Earth.

The square root of 2 also appears in problems of the following nature. Suppose we have a tank of water. We are not given the dimensions of the tank, but we are told that the surface of the water is, say, one foot above the bottom. Suppose the tank has a one-inch diameter drain at the bottom to drain the water off when it is no longer required to be in the tank. Suppose this drainage takes 60 minutes. Now suppose the same tank is again filled with water so that the surface of the water is two feet above the bottom of the tank. In other words, the surface of the water is now twice as high as it was previously. Once again the drain plug is pulled and the tank emptied. How long in this case will it take the tank to empty?

Many people might say it will take twice as long. But that's not correct.

The correct answer is it will take (in minutes) 60 times $\sqrt{2}$. This works out as 84 minutes and 51 seconds. In general the flow of water out of a tank (where it is maintained at a constant height) is proportional to the square root of the water's height. If we let *g* equal the gravitational constant on the surface of the earth (which equals 32.15 feet per second per second) and let *h* equal the height of the water in the tank, it can be proved that the speed of the water flowing from the tank is equal to $\sqrt{2} \cdot \sqrt{gh}$, or, stating it more formally, as below:

$$\text{velocity of water flowing from a tank} = \sqrt{2gh} \text{ or } \sqrt{2}\sqrt{gh}.$$

In other words, if the water flowing from a height of, say, *H* feet takes *n* minutes to empty from a tank, then water flowing from a height of 2*H* feet will take *x* minutes, where:

$$\frac{n}{x} = \sqrt{\frac{H}{2H}}.$$

In the stated problem this equals

$$\frac{n}{x} = \sqrt{\frac{1}{2}}.$$

Squaring both sides of the equation gives

$$\frac{n^2}{x^2} = \frac{1}{2} \text{ or } 2n^2 = x^2.$$

Thus, if *n* equals 60 minutes as in our example, then $2n^2$ equals 7,200 minutes, which equals x^2. The square root of 7,200 is 84.85281 . . . . Thus, in our example, when the tank contains twice as much water as it previously held, it will take 84.85281 . . . minutes for the tank to empty. This result is obtained by the use a well-known theorem in physics known as *Torricelli's law*. The law is named in honor of Evangelista Torricelli, who discovered it in 1643.

The square root of 2 also appears in the following mathematical puzzle. Suppose a column of soldiers 50 yards long is marching forward at a constant rate. The army's mascot dog is at the rear of the army. As the column of soldiers begins to march forward, the dog runs up to the soldier at the front of the line and instantly turns around and runs back toward the last soldier in the column. The dog reaches the last soldier in the column at precisely the point that the line of soldiers has marched 50 yards forward. Assume that the dog's speed is constant. How far does the dog run when the army has marched 50 yards?

There are a number of ways of attacking the problem. Whichever mode of attack is taken, many puzzle enthusiasts are surprised when they learn that $\sqrt{2}$ enters into the solution. The solution is that the dog will run $50 + 50 \cdot \sqrt{2}$ yards, or a total of 120.710678119 . . . yards. In general, if the length of the soldiers is x yards, the dog will run a distance (in yards) equal to x plus x times $\sqrt{2}$.

Here's one simple method of solving the puzzle. Let the distance the dog travels when he reaches the front of the column of soldiers equal $50 + x$. This distance equals $50 + x$. In the same time interval, the column of soldiers will have marched x yards. Therefore, the ratio of the distance traveled by the dog to the distance traveled by the line of soldiers is $\frac{50+x}{x}$.

When the dog is running *toward* the rear of the advancing column of soldiers, he has to run x yards. This time the line of soldiers will have marched $50 - x$ yards. Therefore, the ratio of the distance traveled by the dog to the distance traveled by the line of soldiers is $\frac{x}{50-x}$.

Since the speed of the dog and the soldiers are both constant, we can write:

$$\frac{50+x}{x} = \frac{x}{50-x}.$$

Therefore,

$$50 + x = \frac{x^2}{50-x}.$$

Multiply both sides of the equation by $50 - x$ to obtain

$$2{,}500 - x^2 = x^2$$

or

$$2{,}500 = x^2 + x^2 = 2x^2$$

or

$$1{,}250 = x^2$$

or

$$\sqrt{1{,}250} = x.$$

Therefore, 35.35533906 . . . = x. Thus the total distance traveled by the dog is $50 + x + x$, which equals 120.7106781 . . . yards. Since the line of soldiers marched exactly 50 yards, we find that the dog travels 2.414213562 . . . yards farther than the soldiers. In other words, the dog travels 50 times $(1 + \sqrt{2})$ yards.

Dr. Cong read this chapter and he had a few comments to make concerning $\sqrt{2}$. He sent the following e-mail:

> Hi, Owen,
>
> Thank you for sending a copy of your discussion of the square root of 2.
>
> I'll begin with a little number curiosity. The square root of 2 equals 1.414213562373 . . . . Take the first nine decimals of $\sqrt{2}$ in groups of threes and sum the three numbers:
>
> $$\begin{array}{r} 414 \\ 213 \\ \underline{+562} \\ 1{,}189 \end{array}$$
>
> The number 1,189 is the fifth number such that the square of it produces the fifth number that is both square and triangular (see figure 9.2).
>
> The square root of 2 crops up in many unexpected places in mathematics. Sometimes when solving a mathematical problem involving complex numbers, the problem of finding the square root of i or $-i$ arises. The solutions to both problems involve $\sqrt{2}$. Here are the two square roots of i and $-i$:

$$\sqrt{i} = \frac{1+i}{\sqrt{2}}$$

$$\sqrt{-i} = \frac{1+i}{-\sqrt{2}}$$

We can prove that these are indeed the required square roots by squaring both sides of the two equations. Both squared results should equal i and $-i$. In the first case above, squaring both sides of the equation gives:

$$i = \frac{1+2i+i^2}{2} = \frac{2i}{2} = i\,.$$

Squaring both sides in the second case gives

$$-i = \frac{1+2i+i^2}{-2} = \frac{2i}{-2} = -i\,.$$

I was a little disappointed, Owen, that you did not elaborate on the infinite continued fraction for $\sqrt{2}$ in your chapter. As you pointed out, $\sqrt{2}$ can be expressed as an infinite continued fraction in the following simple manner:

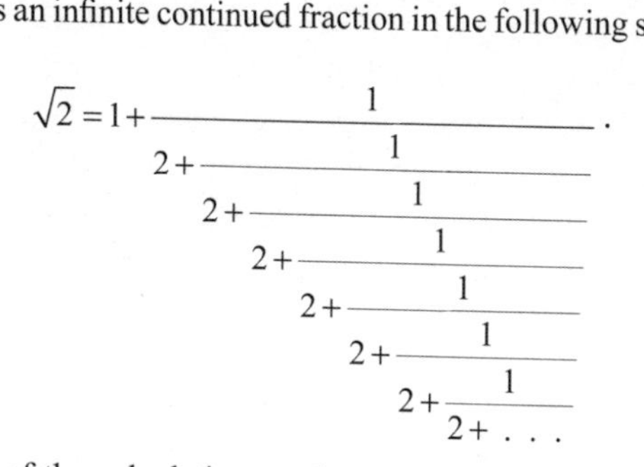

$$\sqrt{2} = 1 + \cfrac{1}{2 + \cfrac{1}{2 + \cfrac{1}{2 + \cfrac{1}{2 + \cfrac{1}{2 + \cfrac{1}{2 + \cfrac{1}{2 + \cdots}}}}}}}\,.$$

Each step of the calculation produces a fraction that approximates the $\sqrt{2}$. The first nine of these fractions are as follows:

$$\frac{1}{1}\quad \frac{3}{2}\quad \frac{7}{5}\quad \frac{17}{12}\quad \frac{41}{29}\quad \frac{99}{70}\quad \frac{239}{169}\quad \frac{577}{408}\quad \frac{1393}{985}\,.$$

If we represent any three consecutive fractions as a/b, c/d, and e/f, we find that

$$e = 2c + a = c + 2d \text{ and } f = 2d + b = c + d.$$

For example, consider the three consecutive fractions 3/2, 7/5, and 17/2. Here $a/b = 3/2$, $c/d = 7/5$, and $e/f = 17/12$. We then obtain the following results:

$$e = 2c + a = c + 2d \text{ and } f = 2d + b = c + d$$
$$17 = 2 \cdot 7 + 3 = 7 + 2 \cdot 5 \text{ and } 12 = 2 \cdot 5 + 2 = 7 + 5$$

If we care to check, we will find the first, third, fifth, seventh, and all the odd-numbered fractions above are all a little less than the true value of $\sqrt{2}$. The second, fourth, sixth, eighth, and all the even-numbered fractions are a little greater than $\sqrt{2}$. This illustrates that there is an unexpected order in this series of fractions.

I notice that you gave the most popular proof that $\sqrt{2}$ cannot be expressed as a fraction. Here's another simple—but lesser known—proof of that fact.

Let a and b be reduced to their lowest form, so that they do not have any common factors. Thus either a or b will be odd. If a/b equals $\sqrt{2}$, then a^2/b^2 equals 2. Thus a^2 equals $2b^2$. The square of any integer contains an even number of prime factors. Therefore, a^2 and b^2 both contain an even number of prime factors. However, $2b^2$ contains one extra prime factor (the number 2). Therefore, $2b^2$ contains an *odd* number of prime factors, and so it cannot equal a number containing an *even* number of prime factors. Therefore, $2b^2$ cannot equal a^2. Therefore, a fraction such as a/b (where a and b are both integers) that equals $\sqrt{2}$ cannot exist. Hence the square root of 2 is an irrational number; that is, it cannot be expressed as a fraction.

Here's a beautiful equation involving $\sqrt{2}$ and the complex numbers i and $-i$. (The number $-i$ appears when we square both sides of the equation in proving that the equation is true.)

$$\sqrt{2} = \frac{\sqrt{i} + i\sqrt{i}}{i}$$

Let us square both sides of the equation. (When we have done so, the left-hand side of the equation will equal 2. Therefore, the right-hand side of the equation will also equal 2.)

Squaring both sides of the equation gives

$$2 = \frac{\sqrt{i} + i\sqrt{i}}{i} \cdot \frac{\sqrt{i} + i\sqrt{i}}{i}.$$

This equals

$$2 = \frac{i + 2i^2 - i}{i^2}.$$

The numbers i and $-i$ cancel out, leaving

$$2 = \frac{2i^2}{i^2}.$$

Thus

$$2i^2 = 2i^2.$$

Divide both sides of the equation by i^2 to obtain

$$2 = 2.$$

Since we deduced this true equation by a number of logical steps from the original equation, that original equation must be true also. In other words, the following equation is true.

$$\sqrt{2} = \frac{\sqrt{i} + i\sqrt{i}}{i}.$$

Who would have thought that the square root of 2 could be expressed in such an elegant manner by the so-called imaginary number, i? The square root of 2, you may recall, is that number that is related to the time for water to flow from a tank if the height of the water level in the tank is doubled; it crops up in the speed of falling bodies toward Earth; it turns up in the ratio of periods of the orbits of planets from a star if one planet is twice the distance from the star as the second planet; and it also crops up in the puzzle about the marching cadets and the trotting mascot dog. We can justly say that all of these phenomena are related to the square root of 2, and all are related to the square root of negative 1. Isn't that amazing?

I found your reference to the octagon and its relationship with $\sqrt{2}$ interesting. Your reference to the puzzle concerning the dog and the soldiers brought back old memories of days long ago when I searched for number patterns in the most unlikely places. I have always searched for number patterns. Sometimes I have found some; other times, well, let's just say I was not so fortunate. I search for number patterns because I believe that a mathematical structure exists out there. Of course, this mathematical structure is not like a physical thing, like a star or a planet or a tree. No, this mathematical structure is an abstract entity. But I do believe it exists. It is, in my not-so-humble opinion, eternal. It has always existed and it will always exist.

In any event, getting back to the puzzle about the dog and the soldiers, I recall reading this version of the puzzle—known as the *courier problem*—in Sam Loyd's *Cyclopedia of Puzzles*, a famous collection of puzzles, originally published in 1914. (Sam Loyd [1841–1911] was a remarkable puzzle genius whose mathematical puzzles and chess puzzles regularly appeared in newspapers in the United States between 1858 and the early twentieth century. Many of his puzzles—both the mathematical and chess variety—were ingenious. For many years the *Cyclopedia* remained out of print and became a collector's item. However, there have been various reprints of this mammoth puzzle volume in recent years.) Loyd also gave a second, much more difficult version of the courier problem in the *Cyclopedia*. I'll discuss that in a moment.

When I was a boy of twelve, I discovered a number of peculiar relationships with the octagon and the puzzle about the marching column of soldiers. I do not have the time to give you all the curiosities I found, so I will give just a few of them. I never before published these peculiarities. Therefore, your readers will be learning of these for the first time, unless they discovered them independently.

If the length of the column of soldiers in the puzzle you mentioned is 50 yards, the dog travels a total distance of 120.7106781 . . . yards when the advancing column of soldiers marches 50 yards. This is equal to 50 times $(1 + \sqrt{2})$. This is the length of the medium diagonal in an octagon that has a side length of 50. In such an octagon, the long diagonal will be 130.656296488 and the short diagonal will be 92.387953251. The ratio between the long and short diagonals is 1.4142135 . . . , which is the square root of 2. (See figure 9.3.)

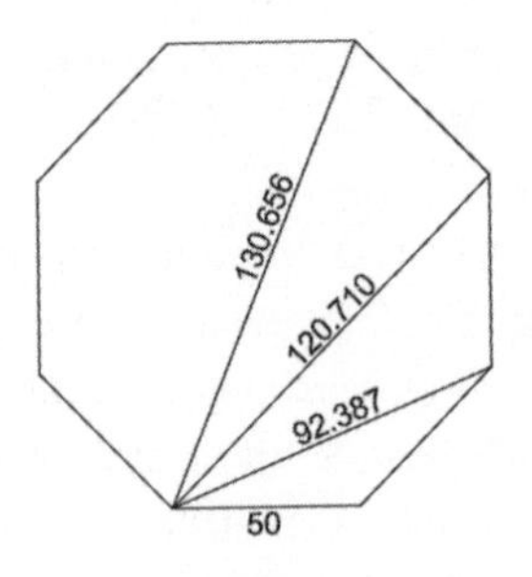

Figure 9.3

If the length of the column of soldiers in the puzzle is 86.5938305 yards, the dog travels a total of 209.0560001 . . . yards. This is the length of the medium diagonal in an octagon that has a side length of 86.5938305. (See figure 9.4.) The length of the long diagonal is 226.2805838 and the length of the short diagonal is 160.0045352 . . . The ratio between these two distances is 1.4142135 . . . , which is the square root of 2.

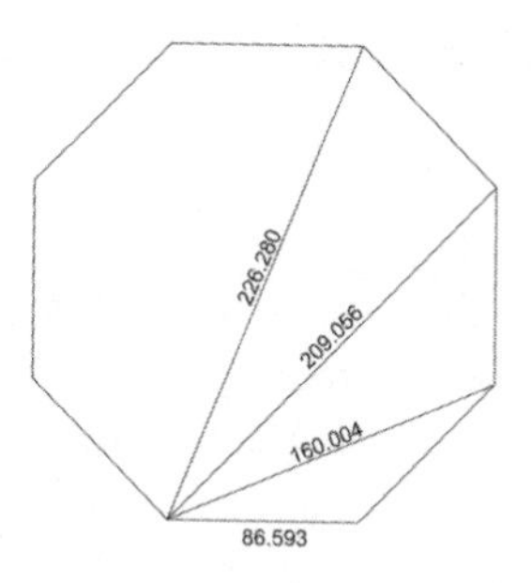

Figure 9.4

That number, 209.056 . . . , is interesting for the following reason. The number 209.056 . . . crops up as the solution to the second version of Loyd's famous puzzle. Here is the second version of the courier problem, or the "marching army problem," as it is sometimes called.

Let's assume that the soldiers in the puzzle were arranged instead in a square formation with a side length of 50 yards. Let us further assume that this square army is advancing forward at a constant rate of speed. Let us assume that the dog begins his journey at the left rear corner of the square army. Let us also assume that the dog travels around this advancing army at a constant rate (first running forward in a parallel direction with the advancing army, then across the front of the marching soldiers, remaining as close to the soldiers as is possible at all times), then down the other side of the advancing army, and finally returns to his starting point (in relation to the marching soldiers) when the army advances exactly 50 yards. The question that arises is, How far did the dog travel?

It is obvious that if the army had remained stationary and the dog ran around it at a constant rate, the dog would travel exactly 200 yards. But of course the army is advancing at a constant rate, so it appears that the dog would have to travel more than 200 yards in order to go around the advancing army. The question is, How do we calculate the distance the dog travels?

The puzzle is not easy to solve. Although Loyd gives the correct answer to both versions of the marching army problem, and hints in the *Cyclopedia* how the first part of the puzzle is solved, he does not supply a method of solution for the first or the second and much more difficult part of the problem. That is regrettable, because many recreational mathematics enthusiasts would have liked to have seen how the famous American puzzle genius would have solved the problem. Perhaps he had a method of solution that has not yet been discovered by others.

The famous recreational mathematician Martin Gardner (1914–2010) gives a solution to the second part of the puzzle in his wonderful book *More Mathematical Puzzles of Sam Loyd* ([New York: Dover, 1960], pp. 103,167, and 168). Gardner's solution is as follows:

Let 1 equal length of the soldiers' line and 1 be also the time it takes the soldiers to march a distance of 1. Thus the soldiers' speed will also be 1. Let x be the total distance traveled by the dog and also his speed. The dog's speed relative to the moving soldiers as the dog moves forward is $x - 1$ and on the backward trip it is $x + 1$. On each of the two diagonal trips, the dog's speed relative to the moving army is $(x - 1) \cdot (x + 1)$ or $\sqrt{x^2 - 1}$.

Relative to the moving army, the dog completes the trip around the soldiers in 1 unit of time. Therefore, we can write:

$$\frac{1}{x-1}+\frac{1}{\sqrt{x^2-1}}+\frac{1}{x+1}+\frac{1}{\sqrt{x^2-1}}=1$$

or

$$\frac{1}{x-1}+\frac{1}{x+1}+\frac{2}{\sqrt{x^2-1}}=1.$$

This equation can be rearranged as

$$x^4-4x^3-2x^2+4x+5=0.$$

This equation involves a power of 4 and is called a *quartic equation*. Quartic equations have at most four solutions. In other words, at most there will four different values for x that will make the equation correct.

There is a mathematical procedure (it is not easy!) to solve quartic equations. The four solutions to this quartic are:

$$x = 4.1811254452$$
$$x = -0.7742319565 + 0.5245449075i$$
$$x = -0.7742319565 - 0.5245449075i$$
$$x = 1.3673384678\ldots$$

The only solution that fits the conditions of the problem is that $x = 4.1811254452\ldots.$ Therefore, the speed of the dog is 4.1811254452 . . . times faster than the square formation of soldiers. Since the square formation of soldiers travels a total of 50 yards, the dog travels $50 \cdot 4.1811254452\ldots$, or a distance of 209.056272 . . . yards.

The interested reader may find it a pleasant exercise to use the solution given to extract the following information relating to this famous problem:

	Dog Travels	**. . . while Army Advances**
Forward Trip	65.7177077 yards	15.7177077 . . . yards
First Diagonal Trip	51.494484 yards	12.315939 . . . yards
Backward Trip	40.349582 . . . yards	9.650412 . . . yards
Second Diagonal Trip	51.494484 yards	12.315939 yards
	Total Distance That Dog Travels	Total Distance That Army Travels
	209.0562 . . . yards	49.99999 . . . yards

I hope that the contents of this little e-mail will be of help to you as you attempt to convey to your readers that $\sqrt{2}$ is a truly beautiful identity.

Best regards,

Dr. Cong

CHAPTER 10

THE SQUARE NUMBERS

The series of numbers that is probably best known to the lay person is the series of natural numbers: 1, 2, 3, 4, 5, 6, 7, 8, 9, 10 . . . . By multiplying each one of these by itself, you obtain the series of perfect square numbers: 1, 4, 9, 16, 25, 36, 49, 64, 81, 100 . . . . The square numbers have their own charm and beauty. For example, the sum of the first n odd numbers, beginning with 1, is equal to n^2. Thus, $1 + 3 = 2^2$; $1 + 3 + 5 = 3^2$; $1 + 3 + 5 + 7 = 4^2$; and so on.

All non-square integers (except zero) have an even number of divisors. For example, the divisors of 10 are 1, 2, 5, and 10 itself. Each prime number has two divisors; itself and 1. On the other hand, all square numbers have an odd number of divisors. The divisor of 1 is 1; the divisors of 4 are 1, 2, and 4; the divisors of 9 are 1, 3, and 9. And so on.

If you multiply a square number by another square number, the answer will always be a square number. Some square numbers, but not all, when added to another square number, equal a square number. The two smallest instance of this in integers are $3^2 + 4^2 = 5^2$ and $5^2 + 12^2 = 13^2$. These are the two smallest cases of the celebrated Pythagorean theorem ($a^2 + b^2 = c^2$).

Finding the digital root is a handy technique to check to see if a number is a perfect square. The digital root of a number is easily obtained as follows: Add the digits contained in the number together. Then add those remaining digits together. Continue the process until only one digit remains. That one final digit is known as the *digital root* of the number being checked. If the digital root is 1, 4, 7, or 9, the number being checked *may* be a square, but is not necessarily square. However, if the digital root is *not* 1, 4, 7, or 9, it definitely is not a square number.

Suppose we want to check if a number, say, 4,489, is a square number. By adding the digits together, we obtain 25. Then, adding the digits of 25 together, we obtain 7. Since the number 4,489 has a digital root of 7, we can conclude that it *may* be a square number. As it happens, it is a square number. It equals 67^2. Suppose we check the digital root of 528 in order to determine if it is a square number. The digital root of 528 is 6. The digital root of a square number cannot be 6, so 528 is definitely not a square number. Suppose we check 16,636 to see if it is a square. The digital root of 16,636 is 4. This tells us that 16,636 *may* be a square number, but not necessarily is one. As it happens, 16,636 is not a square number. Its square root equals 128.9806 . . . .

Some curious facts concerning square numbers have been established over the years. The names of many of those who discovered these properties have been lost in the mists of time, so it is difficult to give credit to the original discoverers. Mathematicians are, by nature, a curious bunch, and if they see something curious about one set of numbers, they usually ask why that curiosity occurs in that particular set and not in other sets. In many cases, this is how discoveries concerning the properties of integers are made. The following are just some of these discoveries.

The product of distinct non-zero integers in arithmetic progression is a *square* number only for –3, – 1, 1, and 3. The product equals 9.

When n equals 1, then $2^n - 1$ equals 1. This is the only occasion when $2^n - 1$ equals a square number. The only time when $2^n + 1$ equals a square number is when n equals 3; $2^n + 1$ then equals 9.

Every square number greater than 1, n^2, has the following identity: $n^2 = n + (n-1)^2 + (n-1)$. Taking 7^2 as an example, you find that $7^2 = 7 + 6^2 + 6$. You might also find that $n^2 = 2(n-1)^2 - (n-2)^2 + 2$. Once again taking 7^2 as an example, you find that $7^2 = 2 \cdot 6^2 - 5^2 + 2$.

The sum of the squares of the natural numbers, beginning with 1, is $1^2 + 2^2 + 3^2 + 4^2 + \ldots$. The resulting sums are 1, 5, 14, 30, 55, 91, 140, 204, 285, 385, 506 . . . These sums can be found by using the following formula:

$$\frac{n(n+1)(2n+1)}{6}.$$

Thus the sum of the first twelve squares equals $(12 \cdot 13 \cdot 25)/6 = 650$.

The formula for the sum of the first n even square numbers is very similar to the above formula:

$$\frac{n(n+1) \cdot (2n+1)}{6} \cdot 4.$$

Thus the sum of the first four *even* squares, 4 + 16 + 36 + 64, is 120. You will notice that this is four times the sum of the first four squares.

The sum of the first n odd squares is equal to

$$\frac{n(2n+1)(2n-1)}{3}.$$

Thus the sum of, say, the first four odd squares, 1 + 9 + 25 + 49, is 84.

All integer squares can be written as $a^2 + 2ab + b^2$, where a and b are two integers, not necessarily different. If we let a and b both equal 1, then the formula gives 4. If we let a equal 1 and b equal 2, then $a^2 + 2ab + b^2$ equals 9. If a equals, say, 13, and b equals, say, 19, then $a^2 + 2ab + b^2$ equals 1,024, which is the square of 32. And so on.

The question now arises, Can the formula $a^2 + ab + b^2$ ever produce a square number? The answer is yes, it can. Let m and n be two different integers, so that m is greater than n. Now let $a = m^2 - n^2$; $b = 2mn + n^2$; and $c = m^2 + mn + n^2$. The formula $a^2 + ab + b^2$ will then produce a square integer. For example, let $m = 2$ and $n = 1$. Then $a = 2^2 - 1^2$; $b = 2 \cdot 2 \cdot 1 + 1^2$; and $c = 2^2 + 2 \cdot 1 + 1^2$; therefore, $a = 3$, $b = 5$, and $c = 7$. The formula then produces the triple 3, 5, 7. If $m = 3$ and $n = 1$, the formula produces $a = 8$, $b = 7$, and $c = 13$. If $m = 3$ and $n = 2$, the formula produces 5, 16, 19. And so on.

Here are the first five such triples:

3, 5, 7
7, 8, 13
5, 16, 19
11, 24, 31
7, 33, 37

To see how these triples fit the bill, so to speak, consider the first triple, 3, 5, and 7. If you let $a = 3$ and $b = 5$, the formula $a^2 + 2ab + b^2$ produces 64, which is the square of 8. However, the formula $a^2 + ab + b^2$ produces 49, which is the square of 7. Consider the second triple, 7, 8, 13. Let $a = 7$ and $b = 8$. Then the formula $a^2 + 2ab + b^2$ produces 225, which is the square of 15. Yet the formula $a^2 + ab + b^2$ produces 169, which is the square of 13. And so on.

The great French mathematician Pierre de Fermat once tackled the following number problem: We are given the set of three integers, 1, 3, and 8.

Note that the product of any two of these integers is 1 less than a perfect square. Can we find a fourth integer and add it to the set such that the product of any two of the four integers is 1 less than a perfect square? Fermat solved the problem: The fourth integer is 120.

Here is a simple method to find sets of four numbers such that the product of any pair of them, plus 1, is a perfect square: Let n be a positive integer, with a value of at least 2, or greater. Let the set of four integers be generated by the following procedure

$$n-1 \quad n+1 \quad 4n \quad 4n\,(4n^2-1)$$

The first five smallest such sets are equal to:

Let n equal	$n-1$	$n+1$	$4n$	$4n\,(4n^2-1)$
2	1	3	8	120
3	2	4	12	420
4	3	5	16	1,008
5	4	6	20	1,980
6	5	7	24	3,432

When n equals 2, the smallest set of four integers is found: 1, 3, 8, and 120. When n equals 3, the next set is found: 2, 4, 12, and 420. The product of any pair of integers, plus 1, in this set equals a perfect square. And so on with each of the other sets. As can be seen from the above formula, there are an infinite number of such sets because n can be any one of an infinite number of integers.

Square numbers have many interesting properties. For instance, consider a number that can be expressed as the sum of two squares. One such number is 5. It equals $1^2 + 2^2$. Another such number is 41. It equals $4^2 + 5^2$. Now if we multiply these two numbers together (5 and 41), the result, 205, will be a number that can be expressed as a sum of two squares in *two* ways. $5 \cdot 41 = 205 = 6^2 + 13^2 = 14^2 + 3^2$. This property always holds. It can be expressed mathematically as $(a^2 + b^2)(c^2 + d^2) = (ac - bd)^2 + (ad + bc)^2 = (ac + bd)^2 + (ad - bc)^2$.

The following curious result is connected to square numbers. The product of any four consecutive integers, plus 1, is always a perfect square. Here are the first five examples of this:

$$1 \cdot 2 \cdot 3 \cdot 4 + 1 = (2 \cdot 3 - 1)^2 = 5^2 = 25$$
$$2 \cdot 3 \cdot 4 \cdot 5 + 1 = (3 \cdot 4 - 1)^2 = 11^2 = 121$$
$$3 \cdot 4 \cdot 5 \cdot 6 + 1 = (4 \cdot 5 - 1)^2 = 19^2 = 361$$
$$4 \cdot 5 \cdot 6 \cdot 7 + 1 = (5 \cdot 6 - 1)^2 = 29^2 = 841$$
$$5 \cdot 6 \cdot 7 \cdot 8 + 1 = (6 \cdot 7 - 1)^2 = 41^2 = 1{,}681$$

Does the pattern continue forever? In other words, is the product of four consecutive integers, plus 1, always a perfect square? The answer is yes. It is a theorem well known to students of recreational mathematics.

However, this property is only a special case of a more general theorem that is not so well known: The product of four numbers in an arithmetic sequence (where the difference between each number is d) plus d raised to the fourth power, is *always* a square number. For example, consider the product of the first four odd numbers: 1, 3, 5, 7. The common difference is 2. Therefore, the theorem tells us that $1 \cdot 3 \cdot 5 \cdot 7 + 2^4$ equals a perfect square (it equals 11^2). Or that $7 \cdot 12 \cdot 17 \cdot 22 + 5^4$ is a perfect square. After checking, we find that this is indeed the case. (It equals 179^2.)

We know that a square number is always of the form $4n$ or $4n + 1$, and never of the form $4n + 3$. Why? Well, the square of an even number, $2n$, is $4n^2$. The square of an odd number, $2n + 1$, is $4n^2 + 4n + 1$, which is a number of the form $4n + 1$. There are no other possibilities, so a square number can never be of the form $4n + 3$.

Consequently, this allows us to derive a number of simple but interesting theorems about square numbers. For instance, the series of numbers 11; 111; 1,111; 11,111; . . . never contains a square number. If such a square number exists, call it x^2, we can write $x^2 = 100m + 11$. This means $x^2 = 4(25 + 2) + 3$. But this means that x^2 is a number of the form $4n + 3$, which we have just proved is impossible. Therefore, there is no square number in the infinites series 11; 111; 1,111; 11,111; . . . .

It can also be easily proved that 5 is the only prime, p, in which the following equation holds: $3p + 1 = n^2$. If this equation holds, then $3p = n^2 - 1$ or $3p = (n + 1)(n - 1)$. Therefore either $(n + 1)$ or $(n - 1)$ equals 3, or either $(n + 1)$ or $(n - 1)$ equals p. If $n + 1$ equals 3, then n equals 2, and $n - 1$ equals 1.

But that means that $3p = (n + 1) \cdot 1$. That means that 1 is a prime, which it is not! Therefore, $n - 1 = 1$ cannot be correct. Consequently, $n - 1 = 3$. Thus $n = 4$ and $n + 1 = 5$. We know as a consequence that the equation $3p = (n + 1)(n - 1)$ is equal to $3 \cdot 5 = 4^2 - 1$.

If p is a prime number, the only values of p that are possible so that $5p + 1$ equals a square number is when p equals 3 or 7.

The expression $n^3 + 3n^2 + n$ is just 1 more than a perfect square, if n equals 1. A recreational mathematician looking at this might wonder if that expression can ever be a perfect square for some integer value of n. Alas, the answer is no. It can never be a perfect square. Why? Let's assume that $n^3 + 3n^2 + n = x^2$. Then $n\,(n^2 + 3n + 1) = x^2$. Since $n^2 + 3n$ is a multiple of n, the greatest common denominator of n and $n^2 + 3n + 1$ is 1. Thus both n and $n^2 + 3n + 1$ are square integers. But $n^2 + 3n + 1$ cannot be the square of an integer, because it lies between the squares of two consecutive integers, namely $(n + 1)^2$, which is $n^2 + 2n + 1$, and $(n + 2)^2$, which is $n^2 + 4n + 4$. Therefore, $n^2 + 3n + 1$ cannot be the square of an integer. Hence the expression $n^3 + 3n^2 + n$ can never equal a perfect square.

It is reasoning like this that helps mathematicians make discoveries in mathematics. Usually the only motivating factor is inquisitiveness.

The only known occurrence of $n^4 + n^3 + n^2 + n + 1$ equaling a perfect square is when n equals -1 or 3. When n equals -1, the above expression equals 1, which is the square of 1. When n equals 3, the above expression equals 121, which is the square of 11.

The only square numbers that contain *only* odd digits are 1 and 9.

The sum of consecutive square numbers, beginning with 1, is related to the theory of probability in a surprising way. To see this relation, choose an integer. Say we choose 5. Add 1 to 5 to obtain 6. The number of ways of choosing 2 objects from 6 items is 15. (This is obtained by multiplying 6 by 5 and dividing the result by 2.) This is usually written mathematically as $C\,(6, 2) = 15$. In other words, the number of combinations of 6 items taken 2 at a time is 15.

We now find the number of ways that 3 objects can be chosen from 6 items, and then we will double the result. This is written mathematically as: $C\,(6, 3) = 20$ and $20 \cdot 2 = 40$. Or we could write it like this: $2 \cdot C(6, 3) = 40$. We now add the two results, 15 and 40, to obtain 55. To see how this is related to the sum of the squares of consecutive integers, beginning with 1, recall that the initial number we chose in this procedure was 5. If we sum the squares of the consecutive integers from 1 to 5, we obtain:

$$1^2+2^2+3^2+4^2+5^2=55.$$

This is always the case. We can write this result as follows:

$$1^2+2^2+\ldots n^2=C(n+1,\,2)+2C(n+1,\,3).$$

The sum of the consecutive square integers from 1 to n may also be expressed as follows:

$$1^2+2^2+3^2+\ldots+n^2=\frac{1}{4}C\,(2n+2,\,3).$$

As an example, let us do the calculation when we make n equal to 7. We can say that the sum of, say, the first seven consecutive square numbers is equal to one quarter of the number of ways of selecting 3 objects from 16. Three objects can be chosen from 16 objects in 560 ways. One quarter of this is 140. Therefore, the sum of the first seven squares is 140.

If we are given a perfect square number, n^2, the next consecutive perfect square number is $2n+1$ greater. For instance, if one is given the perfect square number 169, which is 13^2, the next perfect square number is $169+(2\cdot 13)+1$, which equals 196, which is the square of 14.

Here is another unexpected curiosity. Recall the primitive Pythagorean triples, in which the length of the hypotenuse exceeds the length of the longer leg by one unit. The first five of these primitive Pythagorean triples are

$$3^2+4^2=5^2$$
$$5^2+12^2=13^2$$
$$7^2+24^2=25^2$$
$$9^2+40^2=41^2$$
$$11^2+60^2=61^2$$

Note the length of the hypotenuses. They are, 5, 13, 25, 41, 61 . . . . Let us now write the series of integers beginning with 1, so that there is a group of two digits between 1 and the next square number, 4; then a group of four digits between 4 and the next square number, 9; then a group of six digits between 9 and the next square number, 16; and so on. Following this procedure, we obtain:

1, 2, 3, 4, 5, 6, 7, 8, 9, 10, 11, 12, 13, 14, 15, 16 17, 18, 19 . . .

If we sum the digits in each group between the successive square numbers, we obtain the following:

$$2+3=5=1 \cdot 5$$
$$5+6+7+8=26=2 \cdot 13$$
$$10+11+12+13+14+15=75=3 \cdot 25$$
$$17+18+19+20+21+22+23+24=164=4 \cdot 41$$
$$\ldots$$

Notice that the first group sums to the hypotenuse of the first primitive Pythagorean triple (in which the hypotenuse is one unit longer than the longer leg). The second group sums to twice the hypotenuse of the next such primitive Pythagorean triple. The third group sums to three times the hypotenuse of the next relevant primitive Pythagorean triple. And so on.

Consider the following equation:

$$N=2+2\sqrt{12n^2+1}.$$

One might think that if N is an integer in this equation, it could be any integer. This is not the case, however. If N is an integer, N will be a square number that appears in the following infinite series: 16; 196; 2,704; 37,636; 524,176; 7,300,804; . . . To find these numbers, begin with 16. Subtract 2 from 16 and square the result, obtaining 196. That is the second number in the series. To find the third number, subtract 2 from 196 and square the result, obtaining 37,636. This the third number in the series. And so on.

Or we could write the values of n that will make N an integer. The values of n must be 2; 28; 390; 5,432; 75,658; 1,053,780; . . . . These numbers are found as follows: Beginning with 2, the second number in the series is found by multiplying 2 by 14 and subtracting the previous number: $2 \cdot 14 - 0 = 28$. Then multiply 28 by 14 and subtract the previous term: $28 \cdot 14 - 2 = 390$. The next term is $390 \cdot 14 - 28 = 5{,}432$. And so on.

If n equals 2, then $N=2+2\sqrt{12 \cdot 2^2+1} = 16 = 4^2$. So N is 4^2. If n equals 28, the formula is $N=2+2\sqrt{12 \cdot 28^2+1} = 196 = 14^2$. So N is 14^2. And so on.

The infinite sum of the reciprocals of the square integers, beginning with 1, involves the famous number, π, which equals 3.14159265 . . . .

$$\frac{1}{1^2}+\frac{1}{2^2}+\frac{1}{3^2}+\ .\ .\ .\ +\frac{1}{n^2}\ =\frac{\pi^2}{6}=1.64493406\ .\ .\ .$$

I should mention in passing that the sequence of non-square integers is: 2, 3, 5, 6, 7, 8, 10, 11, 12, 13, 14, 15, 17 . . . . These integers can be obtained by the following formula; $F(n) = \lfloor (\sqrt{n} + 0.5) \rfloor + n$. The expression $\lfloor\ \rfloor$ is known as the *floor function.* It simply means to reduce the value inside the symbols $\lfloor\ \rfloor$ to the nearest integer. For example, letting n equal 1, one finds that $F(1) = \lfloor (\sqrt{1} + 0.5) \rfloor + 1$ equals 2. When $n = 2$, the value of $F(2) = \lfloor (\sqrt{2} + 0.5) \rfloor + 2$ equals 3. When $n = 3$, the value of $F(3) = \lfloor (\sqrt{3} + 0.5) \rfloor + 3$ equals 5. And so on.

There are numerous square numbers with very interesting properties. I only have space to mention one of these. Consider 64. The sum of its digits (10) is one more than a square number (9), and the product of its digits (24) is one less than a square number (25). The number 64 is the only square number that has this property.

Squared quantities turn up constantly in natural phenomena. Most often it is in the law of inverse squares, which will be explained momentarily. One example is in the study of astronomy. In order to make calculations and comparisons easier to understand while solving astronomical problems, astronomers usually describe the earth's distance from the sun as 1 astronomical unit. Here is an example of a problem that arises in astronomy. Suppose that the amount of sunlight received by the earth from the sun is expressed as 1. How much sunlight does Mercury, which is the nearest planet to the sun, receive?

This is a problem in which the law of inverse squares applies. Mercury is about 0.387 astronomical units from the sun. This law states that the source of any physical quantity or intensity, such as light or gravitation, is inversely proportional to the square of the distance from the source of that physical quantity. Since Mercury is about 0.387 times the distance from the sun than the earth is, the amount of sunlight received by Mercury is $(1/0.387) \cdot (1/0.387)$ that of the earth. Now 1 divided by 0.387 equals 2.58397 . . . . The square of this equals 6.6769 . . . . Thus Mercury receives 6.6769 . . . times more sunlight than the earth receives.

Translating this information into watts of radiation per kilometer, one can say that Mercury receives about 9,126 watts of light radiation per kilometer from the sun, while the earth receives 1,367 watts per kilometer. In other words, Mercury receives nearly seven times more sunlight than the earth.

On the opposite side of the spectrum, Mars is about 1.523 *farther* from the sun than the earth is. Therefore, it receives (1/ 1.523) · (1/1.523) *less* sunlight than the earth does. Now 1 divided by 1.523 equals 0.65659 . . . . The square of 0.65659 . . . equals 0.43111 . . . . In other words, Mars receives about 43 percent of the sunlight than Earth does.

There are numerous other physical quantities (besides light) in nature that obey the inverse square law. Sound, radiation, the electric field, and gravity all are subject to it.

The following is a well-known approximation formula in physics: $d = 16t^2$. Here the inverse square law enters the picture again. In this formula, d is the distance an object falls (in feet) from rest, and t is the time (in seconds) it takes the object to fall that distance. Suppose we drop a stone from a resting position at a distance of 16 feet. How long does it take before the stone strikes the ground? Plugging the values into the formula, we obtain

$$16 = 16t^2$$

or

$$\frac{16}{16} = t^2$$

or

$$\frac{1}{1} = t^2.$$

Therefore, $1 = t^2$ and so $1 = t$.

Therefore an object dropped from rest at a height of 16 feet takes 1 second to strike the ground. Similarly, if a stone is dropped a distance of 400 feet, it will take 5 seconds for it to strike the ground. We work this out as follows: We write the formula $d = 16t^2$ and then plug in any values we know. This gives

$$400 = 16t^2$$

or

$$\frac{400}{16} = t^2$$

or

$$25 = t^2.$$

Taking the square root of both sides this gives

$$5 = t.$$

The weight of the stone is irrelevant and we assume here that air resistance is zero. (Air resistance is not zero, but it is negligible in these calculations, and so it can be ignored.)

The squared quantities make an additional appearance in problems such as these! Consider the case where a stone is dropped from a height of 400 feet. The formula above tells us that it will drop 16 feet in the first second. Therefore, in the first second the stone will fall 16/400, or 1/25, the distance to the ground. In the second second, the stone will fall 64 feet. This is 64/400, or 4/25, the distance to the ground. In the third second, the stone will fall 144/400, or 9/25, the distance to the ground. In the fourth second, the stone will fall 256/400, or 16/25, the distance to the ground. Finally in the fifth second, the stone will fall 400/400, or 25/25, the distance to the ground. Note that the numerators and denominators in each fraction are all square numbers!

Of course, Isaac Newton's famous formula for universal gravitation leads to squared quantities. Newton's formula basically states the following: Any two bodies in the universe are attracted by a force that is directly proportional to the size of their masses and inversely proportional to the square of the distance between the two bodies. In other words, Newton's formula states that a force actually *exists* between any two bodies in the universe.

Newton's formula is sufficiently accurate for most practical situations. For instance, it was used by NASA scientists to work out the trajectories required to put men on the moon and it is still used by NASA for space flight today.

However, Newton's formula is superseded by Einstein's theory of general relativity. When very large distances are involved, general relativity gives the more accurate results. Hence, general relativity is accepted as being a "better" picture of reality than Newton's theory of universal gravitation.

Einstein's theory of general relativity states that gravity is not a force in itself. It is a consequence of large bodies such as stars and planets curving space-time. As a result, bodies near a large object such as a star or a planet will follow a curved path because the space it travels through—which is referred to as *space-time* in general relativity—is curved. Hence a planet will move in a curved path around a nearby star.

We still state in our everyday language that a planet orbiting a star is doing so because it is attracted by the star's gravitational attraction, as if there was an actual force existing between the star and the planet. However, it is more accurate to state that the planet is moving through a curved path around the star because the star's mass has curved space-time, and therefore that is the only path that the planet can take.

However, since Newton's formula is accurate in our everyday calculations, we will use his formula here.

One way of understanding the formula is to consider the following: Suppose an apple is dropped from rest at the surface of the earth. The earth's radius is approximately 3,963 miles.

Therefore, the apple's distance from the center of the earth is approximately 3,963 miles. The formula to calculate the speed of an object dropped from rest is 32.174 feet · *t*, where *t* is time in seconds. Thus the previous formula helped us calculate that an apple on the surface of the earth will fall from rest approximately 16 feet in 1 second. This second formula helps us calculate that at the end of that first second, the apple will be accelerating at 32.174 . . . feet per second. At the end of the first second, the apple will be traveling approximately 32 feet per second, which equals 192 inches per second. At the end of the second second, the apple will be accelerating at the rate of 64.3 feet per second. At the end of the third second, the apple will be accelerating at the rate of 96.52 feet per second. And so on.

According to Isaac Newton's theory of universal gravitation, if the distance between two celestial bodies, such as planets, or stars, is doubled, the gravitational attraction between them is 2 · 2 or four times less than it was before the distance was doubled. If the distance between two planets is trebled, the gravitational attraction is 3 · 3 or nine times weaker than it was before the trebling of the distances. If the distance between two planets is quadrupled, the gravitational pull is 4 · 4 or sixteen times weaker than before the distance was quadrupled. And so on.

The moon is about 60 times farther from the center of the earth than the apple is. Therefore, according to Newton, the earth's gravity on the moon should be about 60 · 60 or 3,600 times *weaker* than the earth's gravity at the surface of the earth. Now an apple will fall from rest on the surface of the earth by about 192 inches in the first second of fall. Therefore, the moon should be falling toward the earth by 192 · 1/60 · 1/60 inches per second. This works out as 0.053 inches per second. This is close to one-twentieth of an inch per second. This is, in fact, the case.

The moon does fall toward the earth about one twentieth of an inch per second due to the earth's gravity, according to Newton. In other words, the earth's gravitational attraction bends the moon from going off in a straight line into a curve and sends it orbiting the earth.

Albert Einstein's famous formula for the amount of energy wrapped up in matter also involves a squared quantity. Einstein's formula is $E = mc^2$. In this equation, E equals energy as measured in ergs; m equals the mass of an object (in grams); and c equals the speed of light (in units of centimeters per second). The equation tells us that energy and matter are two forms of the same thing—two sides of one coin, so to speak.

Einstein's astonishing equation can be manipulated to show that there is an enormous amount of energy wrapped up in matter. For example, if we were to convert 1 gram of matter into energy, that amount of energy would keep a 100-watt light bulb running continuously for about 28,538 years.[1]

Squared quantities appear in the period of oscillation of a pendulum. The period of oscillation for a pendulum with a small angle (less than 57 degrees) is given by the following approximation formula, where L is the length of the pendulum in meters; T equals time in seconds; and g is the force of gravity ($g = 9.80665\ldots$ *when* expressed in meters per second) at the earth's surface:

$$T = 2\pi\sqrt{\frac{L}{g}}$$

If the length of the pendulum is 1 meter, the approximation formula equals

$$T = 2\pi\sqrt{\frac{1}{9.806\ldots}} = 2.007089\ldots$$

The formula tells us that a pendulum that is 1 meter in length has a period of oscillation of approximately 2.006409 seconds. The value of π^2 is 9.869604 . . . , and the value of g at the earth's surface is 9.806 . . . . This close approximation of the equality of the two values is a neat coincidence, and it lends itself to the following approximate formula: the square of the time of the pendulum's oscillation in seconds divided by 4 gives an approximation for the length of the pendulum in meters. This can be written as:

$$L = \frac{T^2}{4}$$

For example, suppose we are told that the time of the pendulum's period of oscillation is 2.006409 seconds; what, then, is the length of the pendulum in meters? Plug the appropriate numbers in to the following formula:

$$L = \frac{2.006409^2}{4} = 1.006419$$

$$L = 1.006419 \text{ meters in length}$$

The formula tells us that the pendulum is approximately 1 meter in length.

You may wish to know how the approximation formula that gave us this result is derived.

Begin with following formula, where L is the length of the pendulum in meters; g is the value of gravity at the earth's surface measured in meters per second; and T is time measured in seconds:

$$T = 2\pi \sqrt{\frac{L}{g}}$$

Square both sides of the equation. This gives

$$T^2 = 4\pi^2 \frac{L}{g}.$$

Multiply both sides of the equation by g. This gives

$$T^2 g = 4\pi^2 L.$$

Divide both sides of the equation by $4\pi^2$. This gives

$$\frac{T^2}{4\pi^2}\frac{g}{1} = L.$$

Keeping in mind that the value of π^2 and g are approximately equal, we divide the top and bottom part of the left-hand side of the equation by π^2. This gives

$$\frac{T^2}{4}\frac{1}{1} = L$$

or

$$\frac{T^2}{4} = L.$$

In words, this formula tells us that the square of the time of the period of oscillation of the pendulum in seconds, divided by 4, equals the length of the pendulum.

You can rearrange this equation to find the duration of the period of the pendulum. Simply obtain the square root of both sides of the equation. This gives

$$\frac{T}{2} = \sqrt{L}$$

or

$$T = 2\sqrt{L}.$$

The braking distance of a vehicle also obeys the inverse square law. For example, if the brakes are applied on a car being driven at, say, 30 miles per hour on a dry asphalt road, the car will travel 43.0 feet before it comes to a stop. If the brakes are applied on a car that is traveling twice as fast, 60 miles per hour, the car will travel 172 feet before it comes to a stop. Notice that if the speed is doubled, the braking distance is quadrupled!

This is worth bearing in mind, because the information could save lives. Most drivers are shocked to learn that a car traveling at 40 miles per hour will require 76.46 feet before it comes to a stop, but a car traveling at 80 miles per hour requires four times as much, 305.84 feet, before it comes to a stop.

Squared quantities appear in the most unexpected situations. For instance, the *approximate* distance of the horizon is given by the following formula, where h is the height of the observer in feet above sea level and n is the distance of the horizon in miles:

$$\sqrt{\frac{3h}{2}} = n.$$

Here is how the formula works. Suppose the eyes of an observer on the seashore are 6 feet above sea level. The distance of the horizon in miles is then equal to the square root of $(3 \cdot 6)/2$, or $\sqrt{9}$. Thus the horizon is approximately 3 miles away.

If we wish to see double that distance out to sea, that is, from 3 to 6 miles, we have to *quadruple* our height in feet. In other words, we would have to be 24 feet above sea level to see approximately 6 miles out to sea. If we want to see approximately 12 miles out to sea, we have to be 96 feet above sea level. To see approximately 24 miles out to sea, we must be 384 feet above sea level. And so on.

One of the first computers used in calculations involving square numbers was the Electronic Delay Storage Automatic Calculator (EDSAC). It was built at Cambridge University, England. On May 6, 1949, it performed its first calculation when it calculated the first square numbers from 1 to 100. Note that EDSAC began performing its first calculations in the 49th year of the century. The number 49 is a square number. EDSAC could perform 650 instructions per second. Note that 650 equals $25^2 + 5^2$. May 6, 1949, was the 17,658th day of the century. And 17,658 equals $63^2 + 117^2$. EDSAC stayed in regular service for 9 years. And we know that 9 equals 3^2.

The following is a little curiosity that you may enjoy. Let A = 1, B = 2, C = 3, and so on. The sum of the digits in the two words SQUARE DIGITS is 149. Each of those three digits is a square digit!

I will conclude this chapter with a little puzzle. The solution involves the square integers.

Suppose a pupil in a school notices that there are 50 numbered lockers (in order) in a hallway and that each locker is closed. For no particular reason, the pupil opens all 50 lockers. Then she closes every second locker. When that is done, she opens every third locker if it is closed or closes it if it is open. She does the same thing with every fourth locker, every fifth locker, every sixth locker, and so on (opening it if it is closed and closing it if it is open). When she is finished, how many lockers are open?

Dr. Cong has reviewed this discussion of squares from his home on the West Coast of the United States. He is constantly thinking about mathematics, and number theory in particular. So I thought he may have a few interesting comments to share.

He sent back the following reply.

Hi, Owen,

Thank you for sending your piece on square numbers. It is not bad; not bad at all! I will just add a few titbits of information that your readers may like.

The numbers 13 and 31 contain similar digits. The same applies to their squares: $13^2 = 169$ and $31^2 = 961$. A similar curiosity applies to the numbers 12 and 21: $12^2 = 144$ and $21^2 = 441$.

Here are two unusual square numbers:

$495475^2 = 245\mathit{495475}625$

$971582^2 = 943\mathit{971582}724$

The numbers that are squared on the left-hand side of the equation signs appear in the solution on the right-hand side (note that I italicized them for easier locating).

Consider a square number that can be expressed as $a^2 + b^2$. It can be proved that if $a^2 + b^2$ is evenly divisible by $ab + 1$, then the result is a square integer. The smallest example is when a equals 2 and b equals 8. The expression $a^2 + b^2$ then equals 68, and $ab + 1$equals 17. When 68 is divided by 17, the result is 4, which is a square number. The five smallest solutions, giving 4 as the answer to the division, are as follows: (2, 8), (8, 30), (30, 112), (112, 418), (418, 1,560).

Consider the first solution set: (2, 8). Multiply 8 by 4 and subtract the first number, 2. The answer is 30. That will be the second number in the second set. The second number in the first set (8) is the first number in the second set. Each subsequent set is formed in a similar manner. Such sets will give a solution of 4 when $a^2 + b^2$ is evenly divisible by $ab + 1$.

Here is another curiosity concerning square numbers, Owen. It involves Arsenal, the famous English soccer team based in north London, in the United Kingdom. The first letter of their name is the 1st letter of the alphabet. In the 4th year of this century, at the end of the day on October 16, Arsenal had played 9 games of the new soccer season. They had accumulated 25 points in the Premiership League, and had scored 25 goals. They had gone 49 league games (carried over from the previous soccer season) without defeat. They defeated Aston Villa on October 16, in the 4th year of the century, in a game that consisted of 4 goals. Curiously, all the numbers mentioned above are square numbers!

You mentioned a neat fact about the number 64 in your chapter. The number 64 nearly always reminds me of the great American chess player Bobby Fischer. He was, in the opinion of many chess experts, the greatest chess player (and the most controversial) that ever lived. Unfortunately, later in life he descended into a form of madness.

Bobby Fischer was born on March 9, 1943. He died on January 17, 2008. I find it curious that Fischer—a man who displayed brilliance on the sixty-four squares—died at the age of sixty-four. Bobby Fischer's initials are the 2nd and 6th letters of the alphabet. Of course it should not come as too much of a surprise to learn that 2 raised to the power of 6 equals 64. Bobby Fischer died in 2008. The sum of the integers from 1 to 64 equals 2080. That number contains the same digits as 2008.

The 1943th prime number is 16,831. The reverse of that number is 13,861. The sum of the proper divisors of 13,681 (i.e., sum of all the divisors less than 13,681) is 251. When 251 is multiplied by the sum of its digits, we obtain 2008.

Using the alphabet code that you used, Owen, where A = 1, B = 2, and so on, the sum of the letters in the name FISCHER is 68. Fischer was born on the 68th day of the year. For those who believe in astrology, that makes Fischer a *Pisces*. Thus his zodiac sign is a *fish*. Of course, his surname is similarly pronounced *Fish-er*.

Bobby Fischer won the US Championship in chess in the 1957/58 season when he was just 14 years old. He won the World Championship in chess in 1972 when he was just 29 years old. Consider those two numbers: 14 and 29. The number 14 multiplied by 29 equals 406. The 406th prime number is 2,791. Reverse the digits of that number and we obtain 1972, the year Fischer won the World Championship. Recall that Bobby Fischer was born on the 68th day of the year. Curiously, 68 times 29 equals 1972. Isn't that interesting!

Best wishes,

Dr. Cong

SOLUTION TO THE LOCKER PUZZLE

(1) The only lockers that will be open when the student has completed her procedure are the lockers numbered 1, 4, 9, 16, 25, 36, and 49. These are all the square-numbered lockers!

Each square number has an odd number of divisors while each non-square number has an even number of factors. At the beginning, all the lockers are closed. The procedure involves opening/closing each numbered locker (corresponding to the number of divisors of that particular number.

As an example of what happens, consider the locker numbered 1, which is a square number. The number 1 has only one divisor: itself. The procedure adopted by the pupil means that locker numbered 1 will be opened. From that point on, locker numbered 1 will remain untouched. So locker numbered 1 will remain open.

Next consider what happens with the locker numbered 4. As you know, 4 is a square number. The divisors of 4 are 1, 2, and 4. These are its only divisors. When the pupil first opens all lockers, then locker numbered 4 will be opened. When the pupil decides to close every second locker, the locker numbered 4 will be closed. Then the pupil decides to open every third locker if it is closed and closes every third locker if it is open. While doing this, the locker numbered 4 (which is closed at this point) will be untouched. Then the pupil decides to open every fourth locker if it is closed and closes every fourth locker if it is open. While doing this, the pupil will open the locker numbered 4. From that point on, the locker numbered 4 will remain untouched. In other words, it will remain open.

A similar thing happens for all the lockers that have a square number. Because the locker numbered 9, for instance, has an odd number of divisors, 1, 3, and 9, the procedure adopted by the pupil will open, close, and then open the locker numbered 9. From that point on, locker numbered 9 will remain open.

Consider the locker numbered 16. The divisors of 16 are 1, 2, 4, 8, and 16. It has five divisors. The procedure adopted by the pupil will mean locker 16 is first opened, closed, opened, closed, and then opened. From that point on, locker 16 will remain untouched.

Because each non-square integer has an *even* number of divisors, each non-square numbered locker will end up being closed at the end of the procedure. Because each perfect square number has an *odd* number of factors, each square-numbered locker will end up being open at the end of the procedure.

CHAPTER 11

THE TRIANGULAR NUMBERS

It was the great German mathematician Carl Friedrich Gauss who first proved in 1796 the theorem that every number is (a) a triangular number; (b) the sum of two triangular numbers; or (c) the sum of, at most, three triangular numbers.

Triangular numbers are formed as follows:

$$1 = 1$$
$$1 + 2 = 3$$
$$1 + 2 + 3 = 6$$
$$1 + 2 + 3 + 4 = 10$$
$$1 + 2 + 3 + 4 + 5 = 15$$
$$\ldots$$

Figure 11.1

The series of triangular numbers is 1, 3, 6, 10, 15, 21, 28, 36, 45, 55, 66, 78, 91, 105 . . . They are usually written as follows: The first triangular number is $T_1 = 1$; the second triangular number is $T_2 = 3$; the third triangular number is $T_3 = 6$, and so on. Thus the *n*th triangular number is T_n. The *n*th triangular number is given by the formula $\frac{1}{2}n\,(n + 1)$.

Sometimes the following formula is used when describing triangular numbers: $\frac{1}{2}n\,(n - 1)$. This formula gives the $n - 1th$ triangular number.

Triangular numbers may be illustrated by the diagram shown in figure 11.2.

1 3 6 10 15

Figure 11.2

Any odd number is the difference between two triangular numbers. Consider 5: It is the difference between 6 and 1. Or consider 7: it is the difference between 10 and 3. And so on.

Any multiple of n, where n is a positive integer, is the difference between two positive triangular numbers. Consider 35, for instance. It is a multiple of 5. The 34th triangular number is 595. (It equals $34 \cdot 35)/2$. The 35th triangular number is 630. It equals $(36 \cdot 37)/2$. The difference between 630 and 595 is 35. Or consider 52. It is a multiple of 13. The 51st triangular number is 1,326 and the 52nd triangular number is 1,378. The difference between 1,378 and 1,326 is 52. And so on.

There are an infinite number of triangular numbers that are the sum of two other triangular numbers. For example, 6 equals 3 + 3, 36 equals 15 + 21, and 55 = 10 + 45. The next triangular number that is the sum of two triangular numbers is 120. It equals 15 + 105. How are such numbers found? Let m equal a triangular number. Obviously $\frac{m(m+1)}{2}$ and $\frac{m(m-1)}{2}$ are also triangular numbers.

The following equation holds:

$$\frac{m(m+1)}{2} = m + \frac{m(m-1)}{2}.$$

Since m is a triangular number, the triangular number $\frac{m(m+1)}{2}$ is the sum of two triangular numbers.

For example, 6 is a triangular number. Therefore $(6 \cdot 7)/2 = 21$ and $(6 \cdot 5)/2 = 15$ are also triangular numbers. If we now add 21 and 15 together, we obtain 36, which is also a triangular number.

Triangular numbers and squares are intimately connected in many ways. For example, the sum of two consecutive triangular numbers is a perfect square. Suppose we square two consecutive triangular numbers and add the two results. The answer will also be a triangular number. For example, $1^2 + 3^2 = 10 = T_{4.}$ In other words, $(T_{n-1}{}^2) + (T_n)^2 = T_{(n)}{}^2$.

The following relationship also holds:

$$T_n = n^2 - (n-1)^2 + (n-2)^2 - (n-3)^2 + (n-4)^2.$$

For instance, if n equals 4, you obtain:

$$T_4 = 4^2 - (4-1)^2 + (4-2)^2 - (4-3)^2 + (4-4)^2 = 10.$$

The differences between the squares of consecutive triangular numbers are also intimately connected to the cubes of the integers, commencing with 2^3.

$$3^2 - 1^2 = 2^3$$
$$6^2 - 3^2 = 3^3$$
$$10^2 - 6^2 = 4^3$$
$$15^2 - 10^2 = 5^3$$
$$21^2 - 15^2 = 6^3$$
$$\ldots$$

You may recall the primitive Pythagorean triples, in which the hypotenuse is one unit longer than the long leg. The first five of these primitive Pythagorean triples are:

Primitive Pythagorean Triple	Area of the Triangle
$3^2 + 4^2 = 5^2$	6
$5^2 + 12^2 = 13^2$	30
$7^2 + 24^2 = 25^2$	84
$9^2 + 40^2 = 41^2$	180
$11^2 + 60^2 = 61^2$	330

Now note the following results:

$$T_1 \cdot T_8 = 6^2$$
$$T_2 \cdot T_{24} = 30^2$$
$$T_3 \cdot T_{48} = 84^2$$
$$T_4 \cdot T_{80} = 180^2$$
$$T_5 \cdot T_{120} = 330^2$$
$$\ldots$$

Notice that the products of the triangular numbers equal the squares of the area of the triangles formed by the primitive Pythagorean triples (compare them to the right-hand column of the above table).

Similarly, if we form the following sums with the areas of the above triangles, we obtain the following beautiful pattern:

$$6 + 30 = 36 = (T_3)^2$$
$$6 + 30 + 84 = 120 = (T_{15})^2$$
$$6 + 30 + 84 + 180 = 300 = (T_{24})^2$$
$$6 + 30 + 84 + 180 + 330 = 630 = (T_{35})^2$$
$$\ldots$$

(Note that each of the subscripts of the triangular numbers is 1 less than an odd square number.)

The infinite sum of the reciprocals of the triangular numbers is 2. This is easily proved by a clever little trick. First write the reciprocals of the triangular numbers as follows:

$$\frac{1}{1}+\frac{1}{3}+\frac{1}{6}+\frac{1}{10}+\frac{1}{15}+\frac{1}{21}+\frac{1}{28}+\ .\ .\ .\ .$$

Note that each of these infinite terms can be written as follows:

$$\frac{1}{1}=\frac{2}{1}-\frac{2}{2}$$
$$\frac{1}{3}=\frac{2}{2}-\frac{2}{3}$$
$$\frac{1}{6}=\frac{2}{3}-\frac{2}{4}$$
$$\frac{1}{10}=\frac{2}{4}-\frac{2}{5}$$

And so on.

We now substitute each of these pairs of fractions for each term in the infinite sum of the reciprocals of the triangular numbers as follows:

$$\frac{2}{1}-\frac{2}{2}+\frac{2}{2}-\frac{2}{3}+\frac{2}{3}-\frac{2}{4}+\frac{2}{4}-\frac{2}{5}+\ .\ .\ .\ .$$

Beginning with the second fraction in the first pair of parentheses, we note that the second fraction in every one of the infinite pair of parentheses cancels with the first fraction in the subsequent pair of parentheses. The only fraction that does not cancel is the very first one, 2/1. All the other terms vanish via the canceling process! Therefore, the infinite sum of the triangular numbers is 2.

Triangular numbers are intimately related to *perfect numbers*. Perfect numbers have fascinated number enthusiasts for centuries.

A perfect number is defined to be a positive number equal to the sum of its divisors, except itself. Many of these enthusiasts are surprised when they first learn that all even perfect numbers are triangular numbers. The smallest perfect number is 6, since 6 = 1 + 2 + 3, which are its divisors, excluding 6 itself. Thus the smallest perfect number can be expressed as follows:

$$2\,(2^2 - 1) = 6.$$

The second smallest perfect number is 28. It equals the sum of its divisors, excluding 28: 1 + 2 + 4 + 7 + 14. Thus the second smallest perfect number can be expressed like this:

$$2^2\left(2^3 - 1\right) = 28.$$

The third perfect number is 496; the fourth is 8,128; and the fifth is 33,550,336.

Every even perfect number can be represented as follows:

$$2^{p-1}\left(2^p - 1\right), \text{ where } p \text{ is a prime number.}$$

At the time of writing in 2015, there are forty-eight known perfect numbers. The forty-eighth perfect number, discovered in January 2013, contains over 34 million digits! To be more precise, the forty-eighth perfect number contains exactly 34,850,340 digits. It equals $2^{57,885,160} \cdot (2^{57,885,161} - 1)$.

It is an open question whether the series of perfect numbers continues forever. Most mathematicians believe that it does, but this has not yet been proved.

Perhaps it can never be proved that the number of perfect numbers is infinite. But mathematicians are usually an optimistic bunch. Most of them have a feeling in their bones that mathematical reality is framed by nature in such a way that allows intelligent beings—even those with relatively little intelligence, such as human beings—some way of proving whether or not perfect numbers are infinite.

Over two thousand years ago, Euclid proved that the prime numbers go on forever, even though we cannot literally count and specify every prime. Thus even creatures with our very limited intellectual powers can sometimes discover mathematical facts that involve the concept of infinity. Such discoveries tend to make most mathematicians hopeful that there is a proof lurking "out there" concerning the infinity or otherwise of perfect numbers that at some future time will be discovered.

Every perfect number discovered so far has been even. It is not known if odd perfect numbers exist. In 1991, it was proved that if an odd perfect number exists, it must be greater than 10^{300}. In 2012, Pascal Ochem and Michaël Rao proved that if an odd perfect number exists, it must be greater than $10^{1,500}$.[1]

The only triangular number that is 1 less than a power of 2 is 4,095:

$$2^{12} - 1 = 4,095 = \frac{90 \cdot 91}{2}.$$

Triangular numbers crop up in various branches in mathematics. Here are three examples. First, if we draw a number of random points on a plane, the number of line segments connecting pairs of points is always a triangular number. Consider the following:

Number of Points	Number of Line Segments Connecting Pairs of Points
1	0
2	1
3	3
4	6
5	10
6	15
	. . .

Triangular numbers are also found in the following set of equations:

	Number of Terms	**Number of Non-square Terms**
$(a)^2 = (a)^2$	1	0
$(a+b)^2 = a^2 + 2ab + b^2$	3	1
$(a+b+c)^2 = a^2 + 2ab + 2ac + b^2 + 2bc + c^2$	6	3
. . .		

The reader will notice that the number of terms (including the number of non-square terms) in each result is a triangular number.

Here is the third example. The following pattern involves the number of diagonals on successive polygons. The second column from the right gives the triangular numbers. The number of diagonals of an n-sided polygon is 1 less than the nth – 2 triangular number.

	Number of Sides of Polygon	Number of Diagonals
Triangle	3	$1 - 1 = 0$
Square	4	$3 - 1 = 2$
Pentagon	5	$6 - 1 = 5$
Hexagon	6	$10 - 1 = 9$
Heptagon	7	$15 - 1 = 14$
Octagon	8	$21 - 1 = 20$
. . .		

Incidentally, the following formula, where n equals the number of sides in the polygon, gives the number of diagonals in a polygon:

$$\frac{n(n-3)}{2}.$$

Triangular numbers are found in many beautiful patterns. Consider the following series, for example.

$$1 + 2 = 1T_2$$
$$3 + 4 + 5 = 2T_3$$
$$6 + 7 + 8 + 9 = 3T_4$$
$$10 + 11 + 12 + 13 + 14 = 4T_5$$
$$15 + 16 + 17 + 18 + 19 + 20 = 5T_6$$
$$\ldots$$

(The subscript of each successive triangular number equals the number of terms on the left of the equations above.)

Triangular numbers are intimately connected with the squares of odd numbers, as can be seen from this pattern:

$$8\,T_0 + 1 = 1 = 1^2$$
$$8\,T_1 + 1 = 9 = 3^2$$
$$8\,T_2 + 1 = 25 = 5^2$$
$$8\,T_3 + 1 = 49 = 7^2$$
$$\ldots$$

Occasionally both the *sum* and *difference* of two triangular numbers equals a triangular number. For example, 21 plus 15 equals 36, and 21 – 15 equals 6. Both 36 and 6 are triangular numbers. The same goes for 171 and 105; their sum equals 276 and their difference equals 66, both of which are triangular numbers as well.

The product of three *consecutive* numbers sometimes equals a triangular number. Here are five examples: $1 \cdot 2 \cdot 3 = 6$; $4 \cdot 5 \cdot 6 = 120$; $5 \cdot 6 \cdot 7 = 210$; $9 \cdot 10 \cdot 11 = 990$; $56 \cdot 57 \cdot 58 = 185{,}136$.

The triangular number 120 is the only triangular that is the product of three, four, and five *consecutive* numbers: ($120 = 4 \cdot 5 \cdot 6 = 2 \cdot 3 \cdot 4 \cdot 5 = 1 \cdot 2 \cdot 3 \cdot 4 \cdot 5$). No other triangular number is known to be the product of four or more consecutive numbers. Of course triangular numbers can be the product of two consecutive numbers. Examples of these are $2 \cdot 3 = 6$; $14 \cdot 15 = 210$, and $84 \cdot 85 = 7{,}140$.

The only triangular numbers (besides the trivial case of 1) containing similar digits are 55, 66, and 666. The only triangular number that is also a Fermat number is 3. Fermat numbers are named after the French mathematician Pierre de Fermat (1601–1605). Such numbers are obtained by the following formula, where *n* is either zero or a positive integer:

$$2^{(2^n)} + 1.$$

When n equals 0, the term in parenthesis equals 2^0, which is 1. The formula then equals $2^1 + 1$, which equals 3. Thus 3 is the smallest Fermat number. When n equals 1, the term in parenthesis is 2^1, which equals 2. The formula then equals $2^2 + 1$, which equals 5. Therefore, 5 is the second smallest Fermat number. When n equals 3, the term in parenthesis is 8. The formula then equals $2^8 + 1$, which is 257. The first six numbers in the series of Fermat numbers are 3; 5; 257; 65,537; 4,294,967,297; 18,446,744,073,709,551,617. Only the first five numbers in this series have been proved to be prime. It is conjectured that there are no more prime numbers in the series. This conjecture has not been proved, however.

The only triangular numbers that are found in the Fibonacci sequence are 1, 3, 21, and 55. The only triangular numbers whose squares are also triangular are 1 and 6. The only *known* numbers that are both triangular and factorial are 1, 6, and 120.

Many number enthusiasts have found it curious that each of the numbers in the group 6, 66, and 666 is a triangular number. This property does not hold with any other digit.

The triangular number 136 is curious because each of its digits is a triangular number.

If T is a triangular number, the following numbers are also triangular:

$$\begin{array}{r} 9T+1 \\ 25T+3 \\ 49T+6 \\ 81T+10 \\ 121T+15 \\ 169T+21 \\ \ldots \end{array}$$

Many students of recreational mathematics are surprised when they first encounter the following pattern:

$$1 + 9 = 10 = T_4$$
$$1 + 9 + 9^2 = 91 = T_{13}$$
$$1 + 9 + 9^2 + 9^3 = 820 = T_{40}$$
$$1 + 9 + 9^2 + 9^3 + 9^4 = 7{,}381 = T_{121}$$
$$1 + 9 + 9^2 + 9^3 + 9^4 + 9^5 = 66{,}430 = T_{364}$$
$$\ldots$$

(Note that each successive subscript is three times the previous subscript, plus one.)

The following is also an interesting pattern:

$$T_{10} = T_{5+5} = 55$$
$$T_{100} = T_{50+50} = 5{,}050$$
$$T_{1{,}000} = T_{500+500} = 500{,}500$$
$$\ldots$$

Number enthusiasts have also discovered interesting identities such as the following:

$$T_8 + T_{35} + T_{23} = T_9 + T_{21} + T_{36}$$
$$8 + 35 + 23 = 9 + 21 + 36$$

By coincidence the sum of the subscripts of the triangular numbers on the left-hand side of the equation above are equal to the sum of the subscripts on the right-hand side.

Here is a beautiful pattern involving triangular numbers and fourth powers of integers:

$$2^4 = T_1 + T_5 = T_3 + T_4$$
$$3^4 = T_2 + T_{12} = T_5 + T_{11}$$
$$4^4 = T_{15} + T_{16} = T_{11} + T_{19}$$
$$5^4 = T_{24} + T_{25} = T_{19} + T_{29}$$
$$6^4 = T_{48} + T_{49} = T_{29} + T_{41}$$
$$\ldots$$

Triangular numbers are closely related to *pentagonal numbers* as well. A pentagonal number is a figurate number that represents a pentagon. (A figurate number is a number that can be represented by a regular geometrical arranged group of evenly spaced points. Figurate numbers are sometimes called *polygonal numbers*.) The nth pentagonal number is given by the formula $\frac{1}{2}n(3n - 1)$. The series of pentagonal numbers, commencing with the smallest, is 1, 5, 12, 22, 35, 51, 70, 92, 117, 145, 176, 210, and so on. Each pentagonal number is one-third of a triangular number. The pentagonal numbers, by the way, are equal to the sum of three triangular numbers in a neat way:

$$T_0 + T_0 + T_1 = 1$$
$$T_1 + T_1 + T_2 = 5$$
$$T_2 + T_2 + T_3 = 12$$
$$T_3 + T_3 + T_4 = 22$$
$$T_4 + T_4 + T_5 = 35$$
$$\ldots$$

Numbers that are simultaneously triangular and square are also possible. To find numbers that are both triangular and square, begin with the three numbers 0, 1, and 6. The third number is simply six times the second number, minus the number before that. Continue to produce numbers in this manner. The series formed is 0; 1; 6; 35; 204; 1,189; 6,930; 40,391; and so on. The *squares* of these terms are the series of numbers that are both *square* and *triangular*. These numbers, commencing with 1, are 1; 36; 1,225; 41,616; 1,413,721; 48,024,900; 1,631,432,881; and so on to infinity.

Readers may also find the following curiosity, discovered by Jean Claude Rosa (1946–) interesting. Rosa was a teacher of mathematics to teenagers at a college in Cluny, France. The 132nd triangular number is 132 multiplied by 133, and the result divided by 2. It equals 8,778, which, coincidentally, is palindromic. Similarly, the 143rd triangular is found to be 10,296. The 164th triangular equals 13,530. Believe it or not, these three triangular numbers form a Pythagorean triple! Amazingly, $8{,}778^2 + 10{,}296^2 = 13{,}530^2$. It is the only *known* Pythagorean triple where each of the three sides is a triangular number.

Triangular numbers turn up now and again in mathematical puzzles. Here are three examples.

Suppose Mr. Smith lives in one house on a street where all the houses are numbered from 1 to n. One day Mr. Smith adds up the numbers on all the houses to the left of his house. The next day, he adds up all the numbers on the houses to the right of his house. To his surprise, the two sums are identical. Assuming that there are fewer than 100 houses on the street, what number is Mr. Smith's house? The answer is given at the end of this chapter. (The Indian mathematical genius Ramanujan (1887–1920) is reported to have mentally solved a problem similar to this.[2])

Here's the second puzzle. Suppose there are n people at a party and each one of the partygoers shakes hands once with every other partygoer. How many handshakes take place at the party?[3] The answer is $T_{n-1.}$ Thus if there are, say, twelve people at the party, the total number of handshakes will be $T_{11,}$ which equals (11 · 12)/2, or 66.

Here's the third puzzle. Suppose the British and Irish rugby team, known collectively as the British and Irish Lions, is touring Australia. This team is made up of Britons (that is, English players, Scottish players, and Welsh players) and, of course, Irish players. There are fifteen players on a Rugby team and also a number of substitutes, usually between four and eight. The *rugby squad* is generally the term used to describe the group consisting of players and substitutes.

Let's say that before the first game against Australia, the British and Irish team coach decides to pick the first two players on the team by placing each name of every member of the squad in a hat. He then randomly picks two names from the hat.

Just before he picks the names from the hat, someone asks: "What are the chances that the first two lucky players drawn from the hat will be Britons?"

The captain of the team, who was a lecturer in mathematics, said: "It's an even chance."

How many players, including substitutes, were on the British and Irish Lions rugby team? The problem can be approached as follows: Let the number of Britons in the squad equal m and the total number of players and substitutes in the squad equal n. Then the chance that the first name out of the hat is that of a Briton is m/n. The chance that the second name out of the hat is *also* that of a Briton is $(m - 1)/(n - 1)$. Since we are told that these chances are even, we can write:

$$\frac{m}{n} = \frac{m-1}{n-1} = \frac{1}{2}.$$

This can be rearranged to form the following equation:

$$m^2 - m = \frac{n(n-1)}{2}.$$

This equation has an infinite number of solutions. But the only one that fits the problem's conditions is when m equals 15 and n equals 21.[4]

So there were 15 Britons in the squad of 21 players. The chance that the first two player names picked from the hat will be Britons is (15/21) · (14/20) = 0.5. Thus an everyday situation where someone decides to choose two names from a larger number of names—under certain conditions—involves triangular numbers. Such is the ubiquity of triangular numbers!

The following beautiful pattern involving triangular numbers was discovered by Don Davis, professor of mathematics at Lehigh University, in Bethlehem, Pennsylvania:

$$\begin{aligned}
6 \cdot (1) + 2 &= 2^3 \\
6 \cdot (1 + 3) + 3 &= 3^3 \\
6 \cdot (1 + 3 + 6) + 4 &= 4^3 \\
6 \cdot (1 + 3 + 6 + 10) + 5 &= 5^3 \\
6 \cdot (1 + 3 + 6 + 10 + 15) + 6 &= 6^3 \\
&\ldots
\end{aligned}$$

Here is my own feeble attempt to find a pattern that also involves the triangular numbers and the cubes:

$$\begin{aligned}
(1 + 6 - 3) \cdot 2 - 0 &= 2^3 \\
(3 + 10 - 6) \cdot 4 - 1 &= 3^3 \\
(6 + 15 - 10) \cdot 6 - 2 &= 4^3 \\
(10 + 21 - 15) \cdot 8 - 3 &= 5^3 \\
(15 + 28 - 21) \cdot 10 - 4 &= 6^3 \\
(21 + 36 - 28) \cdot 12 - 5 &= 7^3 \\
&\ldots
\end{aligned}$$

I asked the good Dr. Cong if he had any comments to make concerning triangular numbers. He sent back the following reply by e-mail.

Hi, Owen,

I enjoyed your discussion on triangular numbers. Here are a few more curiosities concerning these amazing integers that you may bring to the attention of your readers.

Suppose we construct right triangles so that the two legs of the triangle are consecutive triangular numbers greater than 1. For example, the two legs of the first five such right triangles are (3, 6,); (10, 15); (21, 28); (36, 45); and (55, 66). The areas of such triangles are $1 \cdot 3^2$; $3 \cdot 5^2$; $6 \cdot 7^2$; $10 \cdot 9^2$; and $15 \cdot 11^2$. Note the appearance of the triangular numbers in order, along with the squares of the odd numbers.

Also, triangular numbers are closely related to the famous puzzle of the crossed ladders. Suppose two crossed ladders are leaning up against opposite sides of a building, so that the tops of each of the ladders coincide with the tops of each of the buildings. Say one building is 40 feet in height and the other building is 65 feet in height. How high above the ground do the ladders cross?

The solution reduces to this: multiply the two heights given and add the two heights given. Then divide the *sum* of the heights into the *product* of the two heights. So 40 times 65 is 2,600. Divide that by (40 + 65), which is 105. The number 2,600 divided by 105 is 24.76190476; so our answer is 24.76190476 feet high, or if you prefer, 24 and 16/21 feet above the ground. Surprisingly, the distance between the buildings is irrelevant in obtaining the height of the crossed ladders.

How are the triangular numbers connected to this?

To see the connection, take the triangular numbers in pairs, commencing with the second and third triangulars, 3 and 6, then 10 and 15, and so on. Now, let us assume that any of these pairs of numbers represent the height of the two buildings. In such cases the height of the crossing point will *always* be an *integer*. Not only that, but the crossing point will always be *twice* the corresponding triangular number. For example, for the smallest solution, the crossing point will be 2, or just twice the smallest triangular number. For the second smallest solution (see the figure below), the crossing point will be 6, or just twice the second smallest triangular number. And so on.

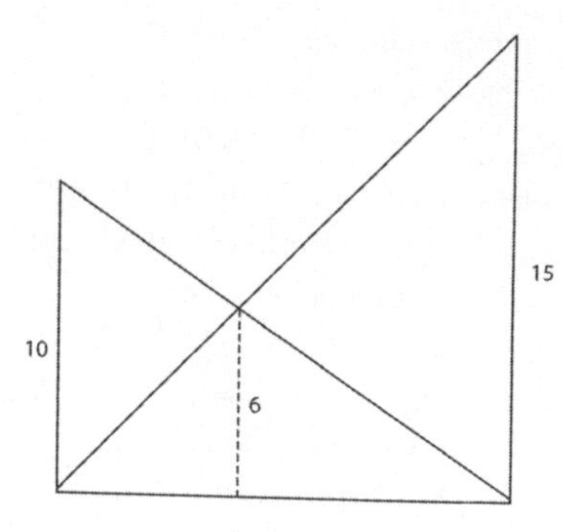

Figure 11.3. Crossing point of ladders is 6 units above the ground.

Consider this table, where the two buildings are designated A and B.

Height of A	Height of B	
3	6	Crossing point of ladders = 2 units above ground
10	15	Crossing point of ladders = 6 units above ground
21	28	Crossing point of ladders = 12 units above ground
36	45	Crossing point of ladders = 20 units above ground
	. . .	

The following pattern involves triangular numbers and the height of the crossing point of the two ladders, when the heights of the building are two consecutive triangular numbers, each greater than 1:

$$3 + 6 - 2 = 7 = 2^3 - 1^3$$
$$10 + 15 - 6 = 19 = 3^3 - 2^3$$
$$21 + 28 - 12 = 37 = 4^3 - 3^3$$
$$36 + 45 - 20 = 61 = 5^3 - 4^3$$
$$55 + 66 - 30 = 91 = 6^3 - 5^3$$
$$\ldots$$

The appearance of those numbers, 7, 19, 37, 61, 91, and so on, is interesting. Why? Because those numbers are the differences between consecutive cubes: $2^3 - 1^3$, $3^3 - 2^3$ . . . are 7, 19, 37, 61, 91 . . . .

Those numbers make an appearance in the following pattern also, where the sum of the squares of the heights of the two relevant ladders is added to the square of the relevant crossing point:

$$3^2 + 6^2 + 2^2 = 7^2$$
$$10^2 + 15^2 + 6^2 = 19^2$$
$$21^2 + 28^2 + 12^2 = 37^2$$
$$36^2 + 45^2 + 20^2 = 61^2$$
$$55^2 + 66^2 + 30^2 = 91^2$$
$$\ldots$$

We can see from these examples that triangular numbers are lurking behind everyday situations, with most people apparently being unaware of their presence. But triangular numbers are found at major public events also. Apparently most people are unaware of their presence there either.

For example, in the finals of major soccer tournaments, such as the FIFA World Cup, we usually see an arrangement of six groups of four teams in the initial stages of these finals. Each of the four teams in each group must play each other once. How many games will take place in each group if there are four teams? The following formula will solve the puzzle: $\frac{1}{2}n(n-1)$. Letting n equal 4, you find that the number of games in each group is 6. In other words, if n is the number of teams in each group, the number of games in each group will be T_{n-1}.

Let any four consecutive triangular numbers equal A, B, C, and D. Numerous interesting properties hold. The following are just two of these:

$$A \cdot D + C \cdot B \text{ is triangular.}$$
$$B \cdot C - A \cdot D = \text{one less than a square number.}$$

Triangular numbers that are palindromic are possible. Examples of these (where the subscripts are also palindromic) are $T_{11} = 66$; $T_{1,111} = 617{,}716$ and $T_{111,111} = 6{,}172{,}882{,}716$.

Three examples of palindromic numbers (where the subscripts are not palindromic) are $T_{109} = 5{,}995$; $T_{173} = 15{,}051$ and $T_{363} = 66{,}066$.

It is curious that the German mathematician Carl Friedrich Gauss discovered the proof that every positive integer is either a triangular number, the sum of two triangular numbers, or the sum of at most three triangular numbers. Why is this curious? Because Gauss was born on the 120th day of the year. The number 120 is, of course, a triangular number.

Furthermore, Gauss was born in the 77th year of the century. I wonder if he was aware of the following curiosities: the 77th prime is 389. Transfer the end digit of that number to the front to obtain 938. Add to that its reversal, and one obtains 1,777. That was the year Gauss was born.

The number 1,777 is the 275th prime number. The digits of 275 sum to 14, and the 14th prime is 43. The number 43 plus the number formed by the reverse of its digits is 77. Gauss was born in the 77th year of the century.

Here is a little curiosity concerning triangular numbers and the life of the US president Abraham Lincoln:

He was born on the 1st day of the week. He was assassinated during the 3rd act of a play, in Ford's Theater, on the 6th day of the week. His successor, Andrew Johnson, was sworn in as president at 10 a.m. the following day. Abraham Lincoln died the morning after being shot, which was the 15th day of the month. His body was boarded onto a train that left Washington, DC, on April 21, heading for the burial site in his hometown.

The numbers mentioned above are the first six triangular numbers in order.

Note that the sum of those triangular numbers, 1, 3, 6, 10, 15, and 21, sum to 56. President Lincoln was 56 years old when he was assassinated. The reverse of the digits of 56 is 65. President Lincoln was assassinated in the 65th year of the century.

I hope you have enjoyed these additional curiosities!

Your friend,

Dr. Cong

SOLUTIONS

(1) (1) Let the house Mr. Smith lives in be m.

This allows us to write:

$$1 + 2 + 3 + 4 + \ldots + m - 1 = m + 1 + m + 2 + m + 3 + m + 4 \ldots + n - 1 + n.$$

Therefore the sum of the numbers of the houses to the left of his house is $m(m - 1)/2$. The sum of the houses to the right of his house is equal to $n(n + 1)/2 - m(m + 1)/2$. We are told that these two values are equal so we can write

$$m(m - 1)/2 = n(n + 1)/2 - m(m + 1)/2$$

or

$$2m^2 = n^2 + n.$$

Now m^2 is obviously a square number, but it is also triangular, since it equals $(n^2 + n)/2$.

We are told that there are fewer than 100 houses on the street. Therefore, m^2 must be less than 100^2, or 10,000. Therefore, all you need to do is to find a square triangular number that when squared is between 1 and 10,000. There is only one such number: 35. That equals m. So m equals 35, and m^2 equals 1,225.

To find the value of n, substitute 35 for m in the following equation:

$$2m^2 = n^2 + n.$$

This gives the quadratic equation $n^2 + n - 2{,}450 = 0$. The two solutions to this equation are –50 or 49. The negative solution does not fit the problem's conditions, so we accept only the positive solution. Therefore, n equals 49. Consequently, there are 49 houses on the street.

Mr. Smith's house number is 35. The sum of all the numbers on the houses to the left of his house is therefore $1 + 2 + 3 + \ldots + 33 + 34 = 595$.

The sum of all the numbers on the houses to the right of his house is $36 + 37 + 38 + \ldots + 48 + 49 = 595$.

CHAPTER 12

THE TRANSCENDENTAL NUMBER KNOWN AS π

It is probably fair to say that the transcendental number best known to the layperson is the number known as π. (A transcendental number is a number that cannot be part of an algebraic expression. I will expand on this point later in this chapter.) The number π is obtained when we divide the circumference of a circle by its diameter. No matter how big the circle, nor what units its diameter or circumference are expressed in, the value of π holds. In other words, π is a dimensionless number.

This ratio was first named π by the Welsh mathematician William Jones in 1706, because π is the first letter of the Greek name for periphery and for perimeter. The symbol π is usually pronounced "pi," which sounds similar to the word *pie*. The great Swiss mathematician Leonhard Euler began using this symbol for the ratio, and the custom spread. However, to this day the word *pi*—instead of the symbol π—is sometimes used to express the ratio.

Many people probably know of π because of its appearance in many geometric shapes. If a circle has diameter d, then the length of the circumference is π times d. Thus if d equals 1, the length of the circumference equals π. If the radius of a circle is r, then the area of that circle is πr^2. Thus a circle with a radius of 1 has an area equal to π. The area of an ellipse is π times the product of the major and minor axes. (The major axis being the longest and the minor axis being the shortest.) The volume of a sphere is $\frac{4}{3}\pi r^3$, where r is its radius. The surface area of a sphere is $4\pi r^2$, where r is its radius. Thus the ratio between the surface area of a sphere and the square of its radius is 4π.

Adults will probably recall from their schooldays that π is often approximated by the fraction $\frac{22}{7}$. That fraction equals 3.14285 . . . . It gives a value for π that is correct to two decimal places. To the first ten decimal places, π is 3.1415926535 . . . . The decimal expansion of π goes on forever.

The infinite digits of π are not without a pattern. They are produced by various mathematical formulae. If we use the correct formulae to calculate π today, and again tomorrow, we will obtain the same series of digits in the same order tomorrow as we obtain today. However, looking at the digits, it *appears* that they are random, that is, that they follow no pattern. No one to date has ever found that the digits of π form a pattern in the sense that given, say, the millionth decimal digit of π, we can predict what the next digit will be.

The number π is an irrational number (which, as we know from previous chapters, means there is no fraction, a/b, where both a and b are integers) that equals π. Other irrational numbers include $\sqrt{2}$, which equals 1.4142135 . . . . The decimal expansion of $\sqrt{2}$ is also never-ending. Although the decimal expansion of $\sqrt{2}$ goes on forever, this number—when multiplied by itself—equals exactly 2. Down through the centuries, some mathematicians wondered if the same thing applies to π. Perhaps, they thought, it also may be the root of some integer?

However, π, or any multiple of it, cannot be the root of any integer. This is because π is not only an irrational number; it is also a *transcendental* number. In other words, π transcends all algebraic equations. Put simply, this means that π cannot be expressed as the root of an algebraic equation. There is no integer that equals a power of π, or a root of π, or a multiple of π.

There are algebraic expressions that come *close* to equaling the true value of π. For instance, consider the equation $9x^4 - 240x^2 + 1{,}492 = 0$. If we substitute π for x in this formula, the right-hand side of the expression equals –0.023236 . . . . The answer may be relatively close to zero, but it is *not* zero. In other words, the equation $9x^4 - 240x^2 + 1{,}492 = 0$ cannot exist if x is equal to π. Nor can any other such algebraic equation.

The number π is found in many famous equations and formulae. For example, consider the Heisenberg uncertainty principle, which is at the heart of quantum mechanics. This uncertainty principle tells us that at its very core, nature behaves in a strange, unpredictable manner. It is impossible to measure the location of a subatomic particle (such as an electron) and its speed simultaneously. Heisenberg's uncertainty principle contains the number π.[1]

The number π is also found in Einstein's field equations. These are ten equations published by Albert Einstein that explain gravitation as the result of spacetime being curved by the presence of matter and energy.[2] Numerous scientific laws, such as Stoke's law, also contain π. This law concerns the frictional force exerted on small spherical objects that are immersed in a viscous fluid.[3]

Many problems in number theory have solutions that involve π. For instance, the probability that two numbers chosen at random are relatively prime is $6/\pi^2$. The probability that an integer does not contain any squared factors is also $6/\pi^2$. The average number of ways to write a positive integer as the sum of two perfect squares (taking the order into account) is $\pi/4$.

The great German mathematician Gottfried Leibniz discovered the following simple expression[4] for π:

$$\pi = 4 \cdot \left(\frac{1}{1} - \frac{1}{3} + \frac{1}{5} - \frac{1}{7} + \frac{1}{9} - \frac{1}{11} + \cdots \right).$$

The English mathematician John Wallis (1616–1703) is said to have discovered the following formula[5] for π in 1655:

$$\pi = 2 \cdot \frac{2 \cdot 2 \cdot 4 \cdot 4 \cdot 6 \cdot 6 \cdot 8 \cdot 8 \cdot}{1 \cdot 3 \cdot 3 \cdot 5 \cdot 5 \cdot 7 \cdot 7 \cdot 9 \cdot 9} \cdots.$$

There are so many beautiful equations that involve π that it is difficult to choose which are the most beautiful.[6] I believe the following is nice:

$$\frac{\pi}{2} = 1 + \frac{1}{3} + \frac{1 \cdot 2}{3 \cdot 5} + \frac{1 \cdot 2 \cdot 3}{3 \cdot 5 \cdot 7} + \frac{1 \cdot 2 \cdot 3 \cdot 4}{3 \cdot 5 \cdot 7 \cdot 9} + \cdots$$

There are many infinite continued fractions involving π as well. It appears that a pattern does not emerge when π is expressed as a *simple* infinite continued fraction. (A *simple* infinite continued fraction contains only 1s in its numerators.) However, beautiful patterns do appear when π is expressed as a *non-simple* infinite continued fraction. Here is a non-simple infinite continued fraction that equals π:

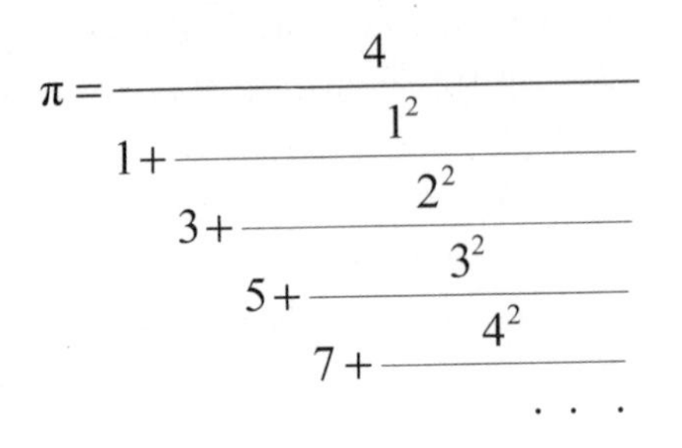

When we look at this continuous fraction, we would never guess that we see the consequences of such a beautiful number every day when we look at circular objects.

The fact that π is an irrational number and that its decimal digits go on forever have led many number enthusiasts to wonder if the decimal digits of π repeat at some point in its infinite decimal expansion. For example, perhaps after a trillion, trillion, trillion, trillion decimal digits, the series 14159265 . . . starts again and continues for another trillion, trillion, trillion, trillion places, at which point the series 14159265 . . . begins again.

This, however, cannot be the case with π. To see why, consider the simple decimal expansion 0.123123123123 . . . , where the digits 123 are repeated in that order forever. Can this simple expression be expressed as a fraction? The answer is yes. To prove this, let $x = 0.123123123123\ldots$. Then $1{,}000x = 123.123123123123\ldots$. Subtracting x from $1{,}000x$ gives $999x = 123$. Therefore, $x = 123/999$. Thus, x can be expressed as a fraction.

Thus, if the digits of π repeated at some point, π could also be expressed as a fraction. But it can be mathematically proved that π cannot be expressed as a fraction. Therefore, the digits of π do not repeat at any point.

The number π crops up in many areas of number theory, too. For instance, consider integers that are the sum of two squares. They have been studied by many of the great mathematicians, including Leonhard Euler. The great Swiss mathematician was aware that 1, for instance, can be expressed as the sum of two squares, ignoring order and signs, in four ways as follows: $+/-1^2 +/-0^2$; 2 can also be expressed as the sum of two squares in four ways: $+/-1^2 +/-1^2$; 3 cannot be expressed as the sum of two squares, but 4 can, in four ways. The number 5 can be expressed as the sum of two squares in eight ways. And so on.

Euler asked himself if there is a limit to how many ways a series of consecutive integers up to some integer N can be expressed as the sum of two squares, where the order and signs of the decomposition into squares is ignored, and where N gets larger and larger. Euler discovered that there is such a limit. That limit is π.[7]

For example, the first 51 integers (counting zero as an integer) can each be expressed as the sum of two squares, ignoring order and signs, in a total of 161 ways. Thus the *average* number of ways these 51 integers can be expressed as the sum of two squares is 161 divided by 51, which equals 3.156862 . . . . This result is not too far away from 3.14159265 . . . , the number known as π. This is not a coincidence. It can be proved that as the number of integers increases from one to infinity, the average number of ways the number of integers can be expressed as the sum of two squares approaches the value of π.

There are many questions linking the digits of π and the prime numbers. Here is a famous one. Consider the initial consecutive digits of π and form numbers from them as follows: 3, 31; 314,159; 31,415,926,535,897,932,384, 626,433,832,795,028,841. Each of those numbers is prime. These primes are known as *pi-primes*. The number of digits in each of these pi-primes is 1, 2, 6, and 38. Three other pi-primes have been discovered to date. They contain 16,208 digits; 47,577 digits; and 78,073 digits, respectively. Are there any more such primes within the digits of π? No one knows!

Euler discovered the following formula that illustrates how π and the primes are intimately related:

$$\frac{\pi}{4} = \frac{3}{4} \cdot \frac{5}{4} \cdot \frac{7}{8} \cdot \frac{11}{12} \cdot \frac{13}{12} \cdot \frac{17}{16} \cdot \;\ldots\; .$$

All the odd prime numbers (in order) appear as numerators on the right-hand side of the equation. The denominators are each a multiple of four that is nearest to the corresponding numerator. Many mathematicians and number enthusiasts find this relationship between π and the primes to be truly beautiful and amazing!

There are many unanswered questions concerning π. For example: Is π a *normal* number? What I mean by this is: Are the decimal digits of π evenly distributed as we would expect in a random sequence of digits? For example, we would expect every single digit to occur one tenth of the time in an infinite random sequence of digits; every two-digit number to appear one hundred times; every three-digit number to appear one thousand times; and so on. The number π has passed every test thus far to check if it is normal, and so it appears from the evidence gathered to date that π is probably a normal number. *Probably* is the key word here. It has not yet been proved if π is normal, although most mathematicians believe it probably is.

Pi crops up in approximate solutions to many problems in probability theory as well. Suppose, for example, that we toss a fair coin an even number of times, say, $2n$ times. What is the probability that exactly 50 percent of the tosses will result in heads and 50 percent will result in tails? Suppose n equals 5 and the coin is tossed ten times. There are precisely 252 ways five objects can be chosen from ten items. Thus there are 252 ways that five coins can be chosen from ten coins.

It is easy to see that if we toss a coin once, there are two outcomes to the way the coin can fall. If we toss it twice, there are 2^2, or 4, possible outcomes. If the coin is tossed three times, there are 2^3, or 8, possible outcomes. And so on.

The total number of ways 10 coins can fall either heads or tails is 2^{10}, or 1,024. Therefore, the chances of getting exactly five heads or five tails if ten coins are tossed are 252/1,024. This equals 0.2460937. . . .

The probability of getting 5 heads and 5 tails in 10 tosses of a fair coin is usually expressed mathematically as follows:

$$\frac{\binom{2 \cdot 5}{5}}{2^{10}} = 0.2460937\ldots = \text{probability of obtaining 5 heads in 10 tosses.}$$

The answer to this probability question (and similar ones) can be approximated. Curiously, π enters the approximation. The *approximate* answer to the probability that in $2n$ tosses of a fair coin exactly n heads or n tails will fall is

$$\frac{1}{\sqrt{n\pi}}.$$

Thus in the problem above the probability is

$$\frac{1}{\sqrt{5\pi}}.$$

This equals 0.25231325 . . . . The true answer is 0.2460937. The approximate answer is about 0.025 percent too much. The percentage error decreases as the value of n increases. Suppose n equals 20. Then the problem asks, what is the probability that in 40 tosses of a fair coin exactly 20 heads and 20 tails will fall?

The formula to solve problems such as this is:

$$\frac{\binom{2 \cdot 20}{20}}{2^{40}} = 0.1253706\ \ldots = \text{probability of obtaining 20 heads.}$$

This probability equals 0.12537068 . . . . That is the true answer.

Let's now calculate the approximate answer. The approximate answer is given by

$$\frac{1}{\sqrt{20\,\pi}}.$$

This equals 0.12615662 . . . . The approximate answer is 0.006 percent too much. This shows that the percentage error is decreasing as the value of n increases.

The number π appears in the following approximation, known as *Stirling's formula*.[8]

$$\sqrt{2\pi n} \cdot \left(\frac{n}{e}\right)^{n}.$$

We can use the formula to get a good approximation of a factorial. Suppose we wish to get an approximation of what factorial 5 is. Factorial 5 is usually written as *5!* It merely means $5 \cdot 4 \cdot 3 \cdot 2 \cdot 1$. Substitute 5 for n in the formula above, and we obtain 118.019 . . . . Factorial 5 equals 120, but the formula gives an approximation that is 98.3493 . . . percent of the true value of factorial 5. If we substitute 7 for n in the formula, the formula gives 4980.395 . . . as an approximation for factorial 7. The true value of factorial 7 is 5,041, so the formula gives an approximation that is 98.797 . . . percent of the true value of factorial 7. As the value of n approaches infinity, the formula gives a result that approaches the true value of the factorial n.

In 1900, the American mathematician Derrick Henry Lehmer (1905–

1991) discovered an excellent approximating formula for calculating the number of primitive Pythagorean triangles with a hypotenuse less than N. The formula is $N/2\pi$.[9]

Let's use the formula to calculate (approximately) how many primitive Pythagorean triangles have a hypotenuse less than 1,000. Substituting 1,000 for N in the above formula yields 159.15494 . . . . So the formula tells us that there are approximately 159 primitive Pythagorean triangles with hypotenuses less than 1,000. There are, in fact, *exactly* 158 such primitive Pythagorean triangles.

The number π crops up in other equations where we would not expect to see it, too. Number enthusiasts, for example, are pleasantly surprised when they first learn that π makes an appearance in an equation involving complex numbers, where i is one of the square roots of negative 1. The equation involves natural logarithms.

The ordinary base ten logarithms are as follows: the logarithm of 10 is 1, since 10 raised to the power of 1 equals 10. The logarithm of 100 is 2, since 10 raised to the power of 2 is 100. Similarly, the logarithm of 1,000 is 3. And so on. However, the natural logarithm is based on the number known as e, which equals 2.71828 . . . . (We will discuss e later in the book.) Thus the natural logarithm of the number 7.38905 . . . is 2, since e raised to the power of 2 is 7.38905 . . . . Similarly, the natural logarithm of 148.41315 . . . is 5, since e raised to the power of 5 is 148.41315 . . . .

You may well ask, why is the number e used as this base of logarithms, and why are they called *natural logarithms*? What is so natural about this base of logarithms? One short answer is as follows: Consider the base ten logarithm of 1.001. It equals 0.000434 . . . . In other words, 10 must be raised to the power of 0.000434 . . . to equal 1.001. There is no apparent connection here between the numbers 1.001 and 0.000434. It seems that that number 0.000434 . . . came right out of left field!

Now consider the natural logarithm of 1.001. It is 0.000999 . . . . This number is very close in value to the decimal part of 1.001. The difference between 0.001 and 0.000999 is 0.000001. In other words, for relatively small quantities such as 1.001, the natural logarithm of the number is very close to the value of the decimal part of the number itself. So if the decimal part of a number such as $1 + x$ (where the decimal part, x, is relatively small), then e raised to the power of x will be very, very close to $1 + x$. This makes the choice of e as the base of the logarithm appear natural. This fact is one of many reasons why the logarithm based on e is called the *natural logarithm*.

Now let's get back to that equation involving π and complex numbers. The equation is: *natural logarithm* of $-1 = i\pi$.

This appearance of π in unexpected places in mathematics—such as the equation above—is why this number is so appealing to both mathematicians and nonmathematicians.

Because π is a ubiquitous number in mathematics, it crops up in many mathematical formulae that involve radians of a circle, instead of degrees. A radian equals the circumference of a circle divided by 2π. Therefore, there are 2π radians in a circle. (This fact makes the use of radians—rather than degrees—*natural.*) Since a circle contains 360 degrees, it is easy to calculate that π radians equal 180 degrees and therefore 1 radian equals $180/\pi$ radians, or 57.2957 radians. As a consequence of this, the sine of a small angle, R, measured in radians, is approximately R. For example, the sine of 0.5 radians is 0.47942 . . . radians.

(As you can see, 0.47942 . . . is very close in value to 0.5. Thus the use of radians seems to be a *natural* choice—rather than degrees—to use in measurements concerning circles. On the other hand, consider the use of degrees in circular (or angular) measure. The sine of 0.5 degrees is 0.0008726 . . . . There is no apparent connection between the two numbers, and so there does not appear to be any *natural* reason to use degrees in measurements relating to circles.)

This fact can be used as the basis of a neat calculator trick. Before doing the trick, ensure that your calculator is in degree mode. Punch a number of 5s (say, six of them) into your calculator and then obtain the reciprocal of the result. The reciprocal approximately equals 1.8^{-6}. Now obtain the sine of this number (in degrees). The number should now be familiar: it is approximately n, accurate to the first six places.

Why does the trick work? The reciprocal of the number consisting of six 5s approximately equals 1.8^{-6}. Let this equal an angle measured in degrees. If we're given an angle in degrees, we can easily convert it to radians by multiplying the angle in degrees by π and dividing the result by 180. Thus 1.8^{-6} multiplied by π and divided by 180 gives a number of radians that is close to the value of $\pi \cdot 10^{-(6+2)}$. Because this is a very small angle measured in radians, the sine of it will be close to $\pi \cdot 10^{-(6+2)}$. Thus it is not surprising that the first six digits of π will appear in your calculator at the conclusion of the trick.

There are many other procedures we could perform on calculators so as to obtain a close approximation of the value of π. The following simple procedure gives an *exact* value of π. Ensure your calculator is in the radian mode. Then sum the arctan of 1, the arctan of 2, and the arctan of 3. The answer is π. (The arctan is the inverse of the tangent function.)

The famous English mathematician John Horton Conway (1937–) once pointed out the following curiosity: Consider the decimal digits of π in blocks of ten at a time. The probability that any one of those blocks of ten digits will contain ten distinct digits is about one in 40,000. Thus in a random set of 40,000 digits you could expect to see a string of ten distinct digits just once. If there were fewer than 40,000 random digits, you would not expect (statistically speaking) to find a set of ten distinct digits. However, commencing at the sixtieth decimal digit of π, we find a string of ten distinct digits. Beginning at the sixtieth decimal digit of π, the series of ten digits is 4592307816. In other words, just sixty decimal digits into the value of π, we see a string of digits that one would only expect to statistically see once by examining 40,000 decimal digits of π.

The decimal expansion of π is full of surprises. For example, consider a number such as 999,999. This number contains six 9s. If we were to examine a series of 1,000 random digits, the probability that we would encounter a string of six digits in a row is 0.000955, which is slightly less than one chance in a thousand. In other words, it is extremely unlikely that we will find six digits in a row of 1,000 random digits.

However, six 9s appear in the decimal expansion of π beginning at the 762nd decimal. This is a surprising result, in the sense that we did not have to go as far as the one-thousandth decimal digit of π. Even if we did go as far as the one-thousandth digit of a random set of one thousand digits, we would expect to see a repeat of six digits only once in every one thousand sets of one thousand digits examined.

Consider the following unusual fact: If we inspected a list of ten million random digits, we could statistically expect that there would be 200 occurrences of five consecutive digits in a row appearing in the list. In the first ten million decimal digits of π, there are *exactly* 200 occurrences of five consecutive digits in a row!

Write the following sequence of the first three odd digits: 135. Then place beside each digit a similar digit to obtain the sequence 113355. Now form a fraction, with the last three digits in the numerator and the first three digits in the denominator. In doing so, we obtain the following fraction: 355/113. This fraction equals 3.14159292 . . . . It gives π correct to six decimal places. The first fraction that gives π correct to seven decimal places is 86,953/27,678. That fraction equals 3.14159260 . . . .

How is π used every day in the real world? There are numerous examples. Here are just two: Electric power is usually delivered to households and businesses in the form of alternating current. (*Alternating current* [AC] is the term used to describe an electric current in which the electric charge periodically reverses direction. On the other hand, *direct current* [DC] is the term used to describe an electric current in which the electric charge goes in one direction only. The mathematical formulae explaining the basis of alternating current include π.[10]

The second example of where π is used every day saves hundreds of thousands of travelers both time and money. When airplanes fly long distance, they usually travel on what is known as a *Great Circle*, which has its center at the center of the earth and its radius is the distance from the center of the earth to the earth's surface. Consequently, a Great Circle is the shortest distance between two points on the surface of the earth. Airplanes fly along these routes in order to save time and fuel. Pi is used in determining the distances of these Great Circles. This information is obviously essential for the airlines of the world so that they know in advance how much fuel is required for each leg of the journey and how long the entire trip will take.[11]

Of course, most long-haul airline passengers are unaware of the essential role the number π plays in getting them from one part of the globe to the other. There are other people in the world however, who are aware of π. Many individuals around the world regularly inspect the decimal digits in π, looking for curiosities or patterns in the never-ending series of digits. I tried my hand at this recently and found the following little curiosity: Commencing at decimal position 28 up to position 36, the following nine digits appear: 279502884. Consider those digits in groups of threes. They are 279, 502, and 884. Their reversals are 972, 205, and 488. Form four addition sums as follows:

Sum of Nine Digits of π,Commencing at the 28th Digit,Where Digits Are Arranged in Groups of Three	Reversals of Numbers
279	972
502	205
884	488
1,665	1,665

Sum from above, with Middle Digit Omitted	Reversal of Numbers
29	92
52	25
84	48
165	165

Figure 12.1

It is curious that the sum of 1,665 should be repeated in the upper table of figure 12.1. Consider the digits of the number 1,665. Note that 65 minus 16 equals a square number and that 65 plus 16 also equals a square number. The first time the sequence "1665" appears in π is at decimal position 9,025, which is also a square number! (It equals the square of 95.) Note that once you remove the middle digits from both the original numbers and the reversal numbers, the sum changes from 1,665 to the curious result of 165 (which is 1,665 with one of its middle digits removed as well!). Examining that second sum, 165, two of its factors are 5 and 33, and the number of primes less than 165 equals 5 plus 33. The number 1,665 equals 150 times 11.1, and the number of primes less than 1,665 equals 150 plus 111.

Many may wonder if those digits in π are trying to tell us something. It is almost certain that they are not! But many will find the appearances of 1,665 and 165 in the above sums to be surprising. Now recall the *lo shu* (discussed at length in chapter 1). Surprisingly, the numbers 1,665 and 165 are connected with the *lo shu*. If we consider the numbers in the horizontal or vertical rows in the *lo shu* as three three-digit numbers, their sum also is 1,665. Their *reversals* also sum to 1,665. To top that, if we consider the numbers in the horizontal or vertical rows in the *lo shu* as three two-digit numbers, omitting the central row or central column, their sum is 165. As a final topper, their reversals are—all together now—165!

Many excellent approximations of π involving fractions have been found. The Indian mathematical genius Ramanujan discovered many approximations of π, such as π *approximately* equals $\frac{9}{5}+\sqrt{\frac{9}{5}}$. This approximation equals 3.1416407 . . . . Ramanujan also discovered the following fraction,[12] which is an excellent approximation: $\pi \approx \sqrt{\sqrt{\frac{2,143}{22}}}$. The approximation equals 3.14159265258 . . . .

This last fraction may be obtained in an unusual manner: Consider the number 1,234. Transposing the first two digits and the last two digits, we obtain 2,143. Let 2,143 be the numerator of a fraction. Now partition 2,143 into two parts as follows: 21:43. Subtract 21 from 43 to obtain 22. Let 22 be the denominator of the same fraction. Then obtain the 4th root of the fraction. We find that $\sqrt{\sqrt{\frac{2{,}143}{22}}}$ is approximately 3.14159265258 . . . . This fraction is accurate to π to about one part in a billion!

Although the endless and nonrepeating sequence of decimal digits in π does not appear to have a pattern, it can be expected that a number of *limited* sequences of digits within the famous transcendental number will produce some curiosities. For example, commencing at decimal position 69, there are ten consecutive *even* digits: 6406286208; commencing at decimal position 411, there are ten consecutive *odd* digits: 5759591953.

Consider the number 9,128,219. It is not only a palindrome, but it is prime also. Such numbers are usually described as *pal-primes*. The next larger pal-prime is 9,136,319. In the decimal expansion of π, commencing at position 9,128,219, the sequence 9,136,319 appears!

When curiosities like these are found within the digits of π, it seems to spur on other digit explorers to make further searches in the hope and belief that there are other major surprises lurking within the digits of this famous number. Indeed, these additional searches have proved fruitful over the years.

For example, there are four *self-locating* numbers within the first one hundred million decimals of π. What do we mean by "self-locating"? Let's look at an example to see. The first decimal of π is 1; in other words, the first decimal of π is self-locating. The second self-locating decimal digit within π is 16,470, which commences at decimal position 16,470. The third one is 44,899, which commences at decimal position 44,899. The last and fourth self-locating digit within the first one hundred million digits of π is 79,873,884, which commences at decimal position 79,873,884.

Many other curiosities have been found in π. We can locate interesting spans of numbers (in a particular order) within the decimal expansion of π. For example, the first twelve digits of π, 314159265358, reappear in π commencing at decimal position 1,142,905,318,634. Similarly, the first eleven digits of the transcendental number *e*, 27182818284, appear in π commencing at decimal position 45,111,908,393. The ascending sequence 0123456789 first appears in π commencing at decimal position 17,387,594,880. And the descending sequence 9876543210 first appears in π commencing at decimal position 21,981,157,633.

Here is a little puzzle that involves π. You will probably find the answer surprising. Assume that the earth is a perfect sphere and that its circumference at the equator is 25,000 miles exactly. Suppose a rope is placed taut around the earth at the equator so that the rope is just touching the earth's surface at all points. Suppose the length of the rope is increased by 2π feet and lifted uniformly off the surface of the earth. How far above the surface of the earth will the rope now be? See the solution at the end of this chapter.

I asked my good friend Dr. Cong if he had any observations on the number π that he might like to pass on to us. He kindly sent the following information.

Hi, Owen,

There are many curiosities involving the digits of π. Consider its first three digits. They are 314. Sum the six permutations of 314 and one obtains 1,776, which is the year that the US Declaration of Independence was signed.

Commencing at the 564,665,206th decimal digit of π, the first string of nine 9s in a row begins. We then have to delve into over 76 million more digits of π to find the next string of nine 9s. That second string commences at the 640,787,382nd digit. Beginning at the 1,943,295,791st digit, the third string of nine 9s appears.

Of course there cannot be an *infinite* string of 9s appearing in the decimal digits of π. Why? If a decimal number ends with an infinite number of 9s, it cannot be an irrational number. The proof is easy. For example, consider the decimal number 0.999999 . . . . This number consists of an infinite number of 9s. Is it irrational? No, it cannot be. Let $x = 0.9999999\ldots$. Then $10x = 9.9999999\ldots$. Subtracting x from $10x$ gives $9x = 9$. Therefore, $x = 1$, which is an integer. Therefore, 0.9999999 . . . cannot be an irrational number. In fact, most people find it surprising and almost counterintuitive that 0.9999999 . . . equals 1.

Here are three little curiosities concerning π: Write the first six digits of π: 314159. Now write them in reverse order: 951413. The difference between these two numbers is 637,254, which is a number that consists of the digits 2 through 7. Of the nine digits, 1, 8, and 9 are missing. Partition the number 637,254 into two parts, so that one has 637 and 254. The sum of the two numbers 637 and 254 equals 891, which is a number consisting of the digits 1, 8, and 9.

Consider the following equation involving the first four digits of π, and the nine digits in order: $3,141 = 1 + 2,345 + 6 + 789$. The digits on the right-hand side of the equation are arranged in four groups. The number of digits in each group in reverse order is 3, 1, 4, 1. Those are the first four digits of π.

The German mathematician Ferdinand von Lindemann (1852–1939), proved in 1882 that π is a transcendental number. The following is a curiosity involving his name, the number of the beast, 666, and π. Using the alphabet code where A = 1, B = 2, C = 3, and so on, the sum of the letters in the name *Ferdinand von Lindemann* equals 212. When 666 is divided by 212, the result is 3.141509 . . . , which is not a bad approximation to π.

The following is a nice way to approximate π to two decimal places by using the digits from 1 to 4 only:

$$\pi \approx 3\frac{2}{14}$$

There are other approximations that include π that are remembered mainly for their curiosity value by a small number of people in the world today. One of these approximations concerns the number of seconds in one year. There are 31,557,600 seconds in a year, assuming a year consists of 365.25 days. This figure of 31,557,600 can be approximated to π times 10 million seconds. This equals 31,415,926 seconds. This approximation is 141,674 seconds (or 1.639 days) too little.

In the United States, the United States Customary Units is a system of measurement that is commonly used. In this system of measurement, there are 3.785411784 . . . liters in a gallon. This approximately equals $3 + \pi/4$ liters ($3 + \pi/4$ equals 3.785398 . . .).

Best regards,
Dr. Cong

SOLUTIONS

(1) The solution to the puzzle about the rope around the earth is often greeted with incredulity by those who solve the puzzle.

In the puzzle, the circumference of the earth is given as 25,000 miles. However, this information is not required to solve the problem.

Let the radius of the earth equal R. Thus the radius of the rope before it is lengthened is also R.

The circumference of the earth is $2\pi R$.

Let the extra radius of the rope (after it has been lengthened) equal r.

When the length of the rope has been increased by 2π feet, the new circumference is $2\pi R + 2\pi$. This allows us to write:

$2\pi R + 2\pi$ = New Circumference

or

$2\pi (R + 1)$ = New Circumference

Divide by 2π to obtain the new radius, r. This gives:

$R + 1 = r$

The new radius is 1 foot more than the previous radius!

In other words, the rope will be 1 foot above the surface of the earth at all points! This will be the case no matter what the size of the sphere is! Every 2π of extra feet in the circumference will add approximately 1 foot to the length of the radius.

CHAPTER 13

THE TRANSCENDENTAL NUMBER *e*

The transcendental number known as π is well known to students of mathematics and laypeople alike. It has been known to mathematicians for thousands of years, although its decimal expansion has only been explored beyond the first fifty or so digits since electronic computers were developed. However, the transcendental number referred to as e is not as well known. It has only been recognized as an extremely important number in mathematics since the early years of the seventeenth century.

A number of European mathematicians came tantalizingly close to discovering e and its importance among numbers. Yet it was not until 1683 that the Swiss mathematician Jacob Bernoulli (1654–1705) realized that a significant number lay between 2 and 3, when he attempted to find the limit of the following function, which arises in calculating compound interest:

$$\left(1+\frac{1}{n}\right)^n \text{ as } n \text{ goes to infinity}.$$

Bernoulli was able to deduce by the binomial theorem that the limit of this function lay somewhere between 2 and 3. This is the first known approximation to the number e in history.[1]

Like π, the number e is a constant that is ubiquitous in mathematics. Mathematicians worldwide believe it is as important as π. The number e is found in the study of biology, chemistry, physics, engineering, and the social sciences.

The number e is an irrational number; hence it cannot be expressed as a fraction. Because it is a transcendental number, e cannot be expressed as the root of a polynomial equation. In plain language, that means that we can multiply e by any number, divide e by any number, raise e to any number, or get the nth root of e, and we will still not obtain an integer. The decimal expansion of e goes on forever, with no apparent pattern in its digits. Like all other transcendental numbers, e can only be expressed as the limit of an infinite series or as the limit of an infinite continued fraction.

The value of e to twenty-five decimal places is: 2.71828182845904523-53602874 . . . . If we memorize e to five decimal places, we will have automatically memorized e to nine decimal places because the four digits 1828 repeat just after the 7 in its decimal expansion. It is believed, by the way, that this repetition of 1828 is purely coincidental.

The letter e was first used to represent the number 2.71828 . . . by Leonard Euler, who made many marvelous discoveries about e. No one knows why Euler chose the letter e, but it is unlikely that he chose it because it is the initial letter of his surname. Euler was a very modest man and certainly not the kind of person to name this extremely important constant after himself. And it is not likely that he chose the letter e because it is the first letter of the word *exponential*. It is known that Euler was at that time already using the letter a to represent a number in his work. It is likely that he chose e because that is the next vowel after a in the alphabet. In any event, the letter e was first publicly used to represent the number 2.71828 . . . in a letter from Euler to the Russian mathematician Christian Goldbach (1690–1764) in 1731.[2]

The number e crops up in areas of mathematics dealing with growth and decay. One way of appreciating the importance of e is to consider the behavior of money left on deposit and earning compound interest.

Suppose we deposit \$100 in a bank and further suppose that the deposit earns 6 percent compound interest per annum. If the deposit is compounded just once per annum, then at the end of the first year the \$100 on deposit becomes 100(1.06) = \$106.00.

Suppose, however, that the bank offered compound interest at 6 percent per annum, but that the compounding would be done quarterly at a rate of a quarter of 6 percent, or 1.5 percent. This appears to be a little more attractive than the previous situation. In such a case, the calculations would be $100(1.06/4)^4 = 100(1.015)^4 = 100(1.06136355\ldots) = \$106.136355\ldots$. With this compounding arrangement we would earn a little more than 13.5 extra cents than the \$106.00 in the previous arrangement. Because of this, we might be inclined to think that the more often the compounding is done, the better the interest return will be. Thus if the compounding was done a great number of times within a specific period, say, a year, we may be inclined to think that massive interest returns will be achieved.

This is not the case, however. If the money was compounded ten million times per annum, at the compound interest rate of 6 percent per annum, the calculations would be: $100(1. + .06/10{,}000{,}000)^{10{,}000{,}000} = 100(1.061836546) = 106.1836546\ldots$). That is less than 5 cents more than if the money was compounded quarterly.

As the number of compoundings per annum increases toward infinity, we can see that the deposit plus interest appears to be approaching some sort of limit. This is the limit that Bernoulli noticed. This is where the number *e* makes its appearance.

Irrespective of how often the money on deposit is compounded, the amount on deposit is, in fact, approaching a limit. This limit is the amount of money on deposit (in this case, $100) multiplied by *e* raised to a power equivalent to the rate of interest per year. This equals $e^{0.06} = 106.1836547\ldots.$

Mathematicians usually state this truth in this way: Let r equal the rate of interest per annum. Let N equal the number of times the compounding is to be done. Then $(1 + r/N)^N$ approaches the limiting value of e^r as N approaches infinity. If we make $r = 1$, and N as large as we like, we have a formula that converges on the number *e*. If, for example, we make N equal to 5,000,000, the formula yields 2.71828155 . . . . If we make N equal to 20,000,000, the formula yields 2.71828176 . . . . And so on.[3]

Mathematicians have discovered a method (known as the *binomial theorem*) in which the formula $(1 + r/N)^N$ may be expanded, giving the following infinite series that converges on the number *e*:

$$e = \frac{1}{0!} + \frac{1}{1!} + \frac{1}{2!} + \frac{1}{3!} + \frac{1}{4!} + \frac{1}{5!} + \ldots.$$

(Recall, the exclamation mark is the factorial sign. $0! = 1$; $1! = 1$; $2! = 1 \cdot 2 = 2$; $3! = 1 \cdot 2 \cdot 3 = 6$; $1 \cdot 2 \cdot 3 \cdot 4 = 24$, and so on.) This infinite series is much more convenient (and quicker) for calculating the decimals of *e* than the formula $(1 + 1/N)^N$. Using an electronic calculator, we can quickly determine the sum of the first eight terms of the above series, yielding the value of 2.7182539 . . . . As more terms of the series are summed, the closer the result will be to 2.718281828459045 . . . .[4]

In fact, the following series converges on e^x, even when x is not an integer.

$$e^x = \frac{x}{0!} + \frac{x}{1!} + \frac{x}{2!} + \frac{x}{3!} + \frac{x}{4!} + \frac{x}{5!} + \ldots.$$

Thus if we make x equal to 2, the above equation equals 7.389056098 . . . . This equals e^2. If x equals 3, the above formula gives 20.08553692 . . . . This equals e^3. If x equals 4.5, the above formula gives 90.0171313 . . . . This equals $e^{4.5}$. When x equals 1, the sum of the series equals e^1, or simply e.

A similar formula to the one for obtaining the value of e can be used to obtain the value of $1/e$, which is sometimes written as e^{-1}. The number e^{-1} equals 0.36787944 . . . . The formula is:

$$e^{-1} = \frac{1}{0!} - \frac{1}{1!} + \frac{1}{2!} - \frac{1}{3!} + \frac{1}{4!} - \frac{1}{5!} + \ldots.$$

In addition, the following formula also applies:

$$\left(1 - \frac{1}{n}\right)^n = e^{-1}, \text{ as } n \text{ approaches infinity.}$$

Leonhard Euler discovered the following infinite continued fraction,[5] which illustrates the beauty, simplicity, and elegance of e:

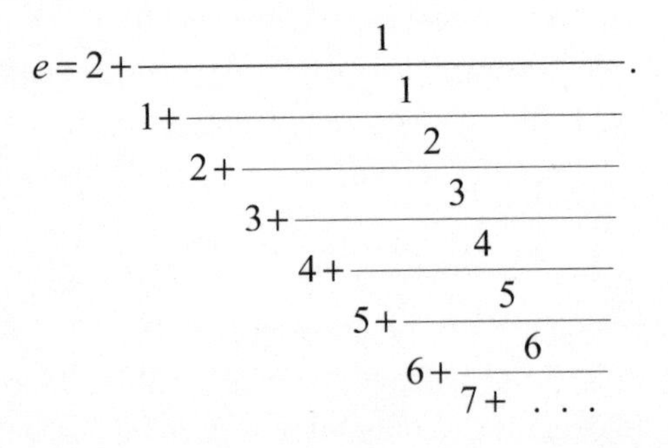

$$e = 2 + \cfrac{1}{1 + \cfrac{1}{2 + \cfrac{2}{3 + \cfrac{3}{4 + \cfrac{4}{5 + \cfrac{5}{6 + \cfrac{6}{7 + \ldots}}}}}}}.$$

In nature, a quantity that *continuously* grows in proportion to the amount of the current quantity is said to be growing at an exponential rate. For example, if a population is growing at a rate of say, 2 percent, it means that at *every instant* the population is increasing by 2 percent of the size of the population at *that instant*.

This facet of nature is what makes the number e so important and why it is found throughout calculus. The reason for this is that e is the only real number that is equal to its own derivative. (A derivative is another word for the *rate of change*.)

Suppose we graph the curve that equals the function $y = x^2$. What this function simply means is that for every point x on the horizontal x axis, the corresponding value of point y on the curve equals x^2. Those beginning to study differential calculus are taught that the function $y = x^2$ is changing at the rate of $2x$ at any point on the curve. In other words, the slope of the tangent at any point on the curve vertically above point x on the x axis is $2x$. Therefore, the derivative of x^2 is $2x$. The function x^3 is changing at the rate of $3x^2$. Therefore, the derivative of x^3 *is* $3x^2$. The function x^4 is changing at the rate of $4x^3$. Therefore, the derivative of x^4 *is* $4x^3$. And so on. As we can see, none of these functions is equal to its own derivative.

Every exponential function has the form Cb^x. When we say that an exponential function is proportional to its own derivative, we are saying that the derivative or rate of change of b^x is proportional to the derivative of b^x. In other words, the derivative of b^x is C times b^x where C is a constant.

What makes the number e unique among numbers is that the function e^x increases at the rate of 1 times e^x. The constant is 1. This is what is meant when we state that the derivative of e^x equals itself. This fact makes e the most convenient choice of base for doing calculus, and it makes e the most suitable base for logarithms; hence the title e: *the base of natural logarithms*.

The crucial fact that is basic to simplifying calculations in calculus is that the derivative of e^x is equal to e^x. It is this astonishing property of the function e^x that makes it ubiquitous in calculus. It can be mathematically proven that e^x is the only positive number with this property.[6]

Let's momentarily digress to see how we can find the value of this special number that is known as e. There are a number of approaches in calculus to find the special number whose derivative is equal to itself. All of these approaches lead to the value of e being equal to 2.71828182845. . . .

The value of this number can be found—without delving into the depths of the calculus—by a method of trial and error. First of all, we note that the derivative of b^x is found by the following procedure:

$$\frac{b^{x+h} - b^x}{h} = \frac{b^x\left(\text{limit of } b^h - 1\right)}{h} = Cb^x$$

Figure 13.1

Here C is a constant, and h is a tiny number. Let's say, for the sake of argument, that h equals 0.000001. What we want to find is a constant, C, equal to 1. Then the derivative of b^x will equal b^x. If x equals 1, then we will find that the value of b is the special number that we are seeking.

The value of b has to be larger than 1. Why? If b equals 1, then b^x will also equal 1, no matter what value x may have. Then $b^h - 1$ will be 0 and so the division by h into 0 results in the right-hand side of the equation equaling 0 also. Similarly, if b is less than 1 (and h is less than 1), $b^h - 1$ will be less than 0 and so the right-hand side of the equation in figure 13.1 will be negative. Thus C is negative. But we are trying to find a solution to the equation so that C equals +1. Therefore, the value of b must be more than 1.

Let the value of b equal, say, 1.5. Let x equal 1, and let h equal a tiny value, say, 0.000001. The formula then gives the constant C a value of 0.40547. We want to increase this constant so that C is equal to 1. If b equals 2, the constant C equals 0.69315. If b equals 2.5, the constant C equals 0.91629. If b equals 2.7, the constant C equals 0.99325. We find that as the value of b is increasing toward 2.7, the procedure (which is equivalent to obtaining the derivative of b^x) is producing a constant, C, that is getting closer and closer to 1.

If b equals 2.8, the constant, C, equals 1.02962. This constant is larger than 1, so b cannot be as large as 2.8. The value of b must equal some number between 2.7 and 2.8. By continuing along this line of reasoning, we find that when b equals 2.718281828 . . . the constant, C, is approaching very quickly the value of 1. So the special value you are seeking for b^x (where x is 1) is that b equals 2.718281828459045 . . . . This number is known universally as e. Its unique property is that the derivative of e^x is equal to e^x.

Exponential equations describe a myriad of natural phenomena, ranging from the growth of populations to modeling the spread of an infection in a population, the spread of cancer cells in a body, the behavior of an avalanche, or the spread of a computer virus around the world, to estimating the number of chain reactions in nuclear fission, and to estimating the rate of radioactive decay.

We can see the effects of the number π in everyday life when we gaze on circles. What about the number *e*? Can we see the effects of that number in everyday life also? Yes, we can.

Consider the catenary curve, for example. (The name *catenary* is derived from the Latin term *catenaria*, which means "chain.") A chain that hangs freely and is supported at two ends, with the only pressure on it being the uniform force of gravity, forms a curve that is known as the catenary. (See figure 13.2.) A spider's web, suspended from fixed points, includes little catenary curves within it. Bees usually make a comb in the shape of a catenary. The cross section of sails bellying in the wind forms a catenary curve that is at a 90-degree angle to the horizontal plane. Surprisingly, the equation for the catenary curve contains the number *e*.[7]

Figure 13.2

Here are some examples of where the number *e* turns up in various situations. Suppose a human body is found in suspicious circumstances.[8] The police are called. A physician will examine the corpse to try and determine the cause and time of death. In determining the time of death, the physician will take the temperature of the body, and will then refer to a chart giving different temperature readings that correspond to the time that has elapsed since the time of death. In calculating those numbers, mathematicians use a formula that involves the natural logarithm, which involves the number *e*. (Note: The symbol used to represent the natural logarithm is *ln*. For example, $e^3 = 20.085536\ldots$. Therefore, the natural logarithm of 20.085536 . . . is 3. This is usually written as $\ln(20.085536\ldots) = 3$.)

Consider sound waves. As sound energy is built up in a room, it increases at a rate that is rapid in the early stages and becomes much slower as time passes. This increase is exponential and is related to the number *e*. When the *source* stops emitting sound, the sound energy—moving away from its source—decays at a rate that is also exponential and is relative to a factor of *e*.[9]

The atmospheric pressure at any given altitude is also related to *e*. At sea level, the atmospheric pressure is 101,325 kilopascals. This is often referred to as a unit of standard atmospheres (atm). Thus at sea level the atmospheric pressure is 1 atm. As one ascends every 25,100 feet of altitude, the atmospheric pressure drops by a factor of $1/e$, which is approximately 0.36787944 . . . . Thus at 25,100 feet, the atmospheric pressure is about 0.37 atm. At an altitude of 1.5 times 25,100 feet, or 37650 feet, the atmospheric pressure drops by a factor of $\frac{1}{e^{1.5}}$, or 1/4.48168907 . . . . Thus at 37,650 feet, the atmospheric pressure is about 0.22 units of atmosphere (atm).[10]

Readers are probably aware of a method of dating ancient objects known as carbon-14 dating. Its official name is *radiocarbon dating*. Every living organism contains carbon-12 and carbon-14. About 5,730 years after an organism dies, the amount of carbon-14 within the organism is reduced to half of what it initially was. When another 5,730 years pass, the amount of carbon-14 is reduced to one quarter of what it initially was. The formula for calculating this rate of decay involves the natural logarithm, *e*, because the rate of decay is exponential. (The reader may recall that the natural logarithm of a number *n* is the value that *e* has to be raised to equal *n*. For example, the natural logarithm of 123 is 4.81218435 . . . , because $e^{4.81218435}$ equals 123. The natural logarithm is usually written as *ln*.) If we let *t* equal the period of time since the death of the living organism; *N* equal the amount of carbon-14 found in the fossil at present; and N_0 equal 100, the following formula[11] can be used to calculate how old the fossil is:

$$t = \left[\frac{\ln\left(\frac{N}{N_0}\right)}{-0.693147} \right] \cdot 5730.$$

Here's an example of how the formula is used. Suppose an archaeologist finds a fossil. It is brought to the laboratory and is tested by scientists. The sample is found to contain 30 percent carbon-14 compared to a living sample. How old is the fossil?

Plugging in the different values, in the appropriate formula gives

$$t = \left[\frac{\ln\left(\frac{30}{100}\right)}{-0.693147} \right] \cdot 5730.$$

This equals

$$t = \left[\frac{-1.20397280\ldots}{-0.693147} \right] \cdot 5730.$$

This in turn equals

$$t = 1.7369660\ldots 5730 = 9952.8154\ldots.$$

This tells us that the living organism died 9,952 years ago. Thus the fossil is 9,952 years old.

The number *e* turns up in the natural world when one considers population growth. Suppose the number of bacteria in a jar at a given time is known to be 200. It is known that the number of bacteria increases by 30 percent every 60 minutes. After 24 hours, how many bacteria are in the jar?[12]

The formula to solve this problem and similar problems is much like that used for calculating compound interest:

$$200 \cdot e^{(0.3 \cdot 24)} = 200 \cdot e^{7.2} = 200 \cdot 1339.43076\ldots.$$

The number of bacteria after 24 hours will be 200 · 1339.43076, or 267,886.

What applies to the growing population of bacteria applies also to the growing population of the human race. For decades, the exponential growth of the world's population and the problems it creates has occupied some of the greatest minds on Earth. Of course, the number *e* makes its appearance in analyzing these problems. Here's an example:

According to an article in the prestigious magazine *Scientific American* (published on March 1, 2007), the world's population reached 6.2 billion people in 2002, and is increasing in the early years of the millennium at a rate of 1.2 percent per annum.[13] At that rate, how many people will be on the planet in the year 2050 (which is 48 years after 2002)? The problem is solved by using the following formula:

$$6.2 \cdot e^{(0.012 \cdot 48)} = 6.2 \cdot e^{0.576} = 11.0292 \text{ billion.}$$

If instead the population increased at the rate of 1 percent per annum, how many people would there be on Earth in 2050? The formula gives

$$6.2 \cdot e^{(0.01 \cdot 48)} = 6.2 \cdot e^{0.48} = 10.0196 \text{ billion people.}$$

In other words, if the rate of increase is 1 percent per annum instead of 1.2 percent, there would be 1 billion people fewer on the planet in 2050. That small reduction in the rate of exponential growth over 48 years makes a huge difference. (It also means that exponential growth, however small, is unsustainable over time, because exponential growth means increases, and increases, and increases . . . , until some other factor inevitably steps in to stop it. In the case of human population growth, that factor will inevitably be famine and disease. It is surely better—and relatively painless—that the human race consciously decides to cut population growth now rather than to have nature do it for us at some time in the future, which will not be painless.) The number *e* helps us to calculate these figures relating to exponential growth relatively easily and quickly.

The number *e* crops up everywhere in mathematics. For example, the sum of the harmonic series (the reciprocals of all integers up to any fraction, $1/n$) is given by the following formula:

$$\frac{1}{1}+\frac{1}{2}+\frac{1}{3}+\frac{1}{4}+\frac{1}{5}+\ .\ .\ .\ +\frac{1}{n} \approx \text{natural logarithm of } n, \text{ for large values of } n\,.$$

The harmonic series is one of the most famous series in mathematics. The sum of the harmonic series diverges. In other words, the sum gets bigger and bigger as more and more terms are added. But the series increases exceptionally slowly! One has to sum more than the first 10^{43} terms to get past 100.

The number *e* is closely associated with the sum of this series. Suppose we want to find out what the first million terms of this series will add up to. (Using advanced mathematics, mathematicians have devised a method that involves a number known as the *Euler-Mascheroni constant* that allows us to calculate this sum relatively quickly and easily. The Euler-Mascheroni constant equals 0.5772157. . . .) The sum of the first million terms of the harmonic series can be obtained by obtaining the natural logarithm of one million. If we have an electronic calculator, we can easily do this. The ln of 1,000,000 is 13.81551056 . . . . In other words, $e^{13.8155105} = 1{,}000{,}000$. Now add the Euler-Mascheroni constant (0.5772157. . . .) to 13.8155105 . . . . The result is 14.3927262. . . .

This answer tells us that despite adding the first million terms of the harmonic series, we still would not have surpassed the number 15.[14]

The following is another famous series that involves the natural logarithm: $\ln(1 + x)$ approximately equals x, when x is small. For example, if x is equal to 0.000001, we obtain the following: $1 + x = 1.000001$ and the ln(1.000001) = 0.00000099999 . . . . The two values are extremely close. As the value of x approaches zero, the natural logarithm of $(1 + x)/x$ approaches the value of 1.[15]

The following is yet another famous series that involves the natural logarithm:

$$\frac{1}{1}-\frac{x^2}{2}+\frac{x^3}{3}- \quad = \ln(x+1),\ \text{when} -1 \text{ is less than } x \text{ and } x \text{ is less than or equal to } 1.$$

If we make x equal to 1, the sum of the above series gives the ln(1 + 1), or the ln(2):

$$\frac{1}{1}-\frac{1}{2}+\frac{1}{3}-\frac{1}{4}+\frac{1}{5}- \quad . . . + \frac{1}{n} = \text{natural logarithm of 2, as } n \text{ approaches infinity.}$$

The natural logarithm of 2 is 0.6931471 . . . . ($e^{0.6931471} = 2$.) The first ten terms of the above sum give 0.6456349 . . . . As we go further into the series, we obtain a sum that gets closer and closer to 0.6931471. . . .

The number *e* crops up in number theory also. Gauss discovered that as we go up the numbers, the quantity of primes less than N, as N approaches infinity, approximately equals N divided by the ln of N. This result can be used to approximate the number of primes less than N. For example, suppose we want to know approximately how many primes are less than 1,000? Divide 1,000 by the natural logarithm of 1,000. The ln of 1000 is 6.9077552 because $e^{6.9077552}$ equals 1,000. The number 1,000 divided by 6.9077552 equals 144.7648. This tells us that there are approximately 144 primes less than 1,000. Actually, there are 168 primes less than 1000. The percentage error is close to 14.2 percent. As we go up the numbers, the percentage error that the formula produces decreases. For example, the formula states that there are approximately 72,382 primes less than one million, while in fact there are 78,498 primes less than one million; the percentage error now is close to 7.7 percent.[16]

The following is known about the number *e* raised to different powers: e^{π} is a transcendental number. At least one of e^e or *e* raised to the power of e^2 is transcendental. It is not known if π^e is transcendental.

When I had completed this chapter, I asked Dr. Cong for any comments he may wish to make in relation to it. He sent me the following e-mail.

Hi, Owen,

I enjoyed your chapter on *e*. Here are a few more odds and ends relating to *e* that your readers may enjoy.

A well-known puzzle in recreational mathematics is known as the problem of the misplaced hats. Suppose five men went to a hotel for a night's carousing. When they arrive, they give their hats to the hat-check girl. All of the men end up having a little too much to drink, and so does the hat-check girl! At the end of the night, the hat-check girl takes each of the five hats at random and gives them back to the five men. What is the probability that all five men got back the wrong hat?[17]

There is a mathematical procedure to solve problems such as this. However, here is an easier way of calculating the probabilities. In this version of the problem (where there are five men), all we need to do is divide five factorial by the transcendental number *e*. The answer to the nearest integer is your solution! So with five men, calculate five factorial. Five factorial equals $5 \cdot 4 \cdot 3 \cdot 2 \cdot 1$, which equals 120. When 120 is divided by the value of *e* (2.718281828), it equals 44.1455. The nearest integer is 44. So the probability that each man will *not* get his own hat back is 44/120. This equals 0.3666666 . . . . This result is close to $1/e$, which equals 0.36787944 . . . .

Suppose there are six men with this hat problem. To find out the probability that each of the six men will not get his own hat back, do the following. Calculate six factorial, which is $6 \cdot 5 \cdot 4 \cdot 3 \cdot 2 \cdot 1$. This equals 720. Divide 720 by 2.718281828 to obtain 264.873197 . . . . The nearest integer is 265. Therefore, the probability that the six men each will not receive his own hat back is 265/720. This fraction equals 0.3680555 . . . . This result is even closer to $1/e$ than the previous result was.

This is no coincidence. It is a curious feature to the problem of the mixed-up hats. When the number of men increases above six or seven, the probability that each man gets his own hat back is virtually unchanged! The probability is very close to $1/e$, which equals 0.36787944 . . . , whether there are 10 men, 10 million men, or 10 billion men!

Here is a similar problem: Suppose we decide to play a lottery game. Let's say that to win the jackpot we have to correctly choose six numbers from a total of forty-five numbers. The number of different ways we can do this is 8,145,060. Now suppose that *exactly* 8,145,060 tickets are sold. What are the odds that at least one person will win the lottery? The answer approaches $1 - (1/e)$, which equals about 63 percent.[18]

The number e is intimately connected to Pythagorean triangles, too. The number of Pythagorean triangles whose perimeter is less than N is approximately $N \cdot \ln 2/\pi^2$. Here the logarithm used is the natural logarithm, based on the transcendental number known as e. The natural logarithm of 2 is 0.6931471 . . . (since the value of $e^{0.6931471\ldots}$ equals 2). The above formula, which is due to American mathematician Derrick Henry Lehmer (1905–1991) is strikingly accurate. For example, using the formula to find the approximate number of Pythagorean triangles with a perimeter less than 10,000 gives $10{,}000 \cdot 0.6931471/\pi^2$, which equals 702.3048 . . . . The actual number of Pythagorean triangles with a perimeter less than 10,000 is 703.[19]

It is extraordinary that the number e is intimately connected to prime numbers. As we go up the numbers toward infinity, the distribution of the prime numbers are marching in step to the tune of e.

The following asymptotic formula flows from the prime number theorem: $p_n \approx n \ln(n)$. (An *asymptotic formula* is a formula that expresses that two quantities that are not equal approach equality as the value of the variable, n, approaches some particular value, usually infinity.) The formula is easy to use. Suppose we want to approximately calculate the one hundredth prime number. To do so, we multiply the ln of 100 by 100. The ln of 100 equals 4.60517018. What this simply means is that 100 equals the number e raised to the power of 4.60517018 . . . . Therefore, to find approximately the one hundredth prime number, multiply 4.60517018 . . . by 100. The answer is 460. Therefore, we know that the one hundredth prime is somewhere in the region of 460. The one hundredth prime number is actually 541. The percentage error is 14.9+ percent.

Using the formula again, we find that the one thousandth prime number is approximately ln 1,000 multiplied by 1,000, which approximately equals 6,907. The one thousandth prime number actually is 7,919. The percentage error here is 12.7+ percent. Using the formula once more, we find that the one millionth prime number is approximately ln 1,000,000 multiplied by

1,000,000, which approximately equals 13,815,510. The one millionth prime number is 15,485,863. The percentage error in this case is 10.7+ percent.

As we go up the numbers, we find that the percentage error is decreasing. In other words, the formula gives an approximate answer that gets closer and closer (in percentage terms) to the true result!

We can get an even better approximation by using the following extension to the above formula: $p_n \approx n \ln(n + n [\ln (\ln (n)\text{-}1)])$[20]

Let's use this formula to approximate the one millionth prime. First, we note that the ln of 1,000,000 is 13.8155105 and that the ln of 13.8155105 is 2.62579191 . . . . The formula is

$$p_n \approx 1{,}000{,}000 \ln 1{,}000{,}000 + 1{,}000{,}000\ [\ln (\ln (n) - 1]$$
$$= 1{,}000{,}000\text{th prime} = 1{,}000{,}000 \cdot 13.8155105 \ldots + 1{,}000{,}000 \cdot [\ln(13.8155105 - 1)]$$
$$1{,}000{,}000\text{th prime} = 1{,}000{,}000 \cdot 13.8155105 \ldots + 1{,}000{,}000 \cdot [(2.62579191 - 1)] = 15{,}441{,}302.$$

The one millionth prime number is 15,485,863. The approximation gives a result that is 99.71 percent accurate.

Here is another instance where *e* plays an unexpected role. Suppose we choose numbers randomly between 0 and 1 until the sum of those numbers exceeds 1. What is the expected value of the number of random numbers required to achieve this? We will probably initially reason that at least two random numbers are required so that their sum will exceed 1. Giving the matter some more thought, we may believe that on average three random numbers are required to yield a sum larger than 1.

Eventually we may be convinced that the number of random numbers chosen so that their sum will exceed 1 probably lies between 2 and 3. This line of thinking leads approximately to the correct answer. The expected number of random numbers chosen between 0 and 1 required so that their sum exceeds 1 is 2.71828182849 . . . . In other words, the answer to the problem is *e*.[21] If we choose numbers from the interval 0, 1, until the sum of those numbers exceeds 2, the expected value of the number of random numbers required is $e^2 - e$.[22]

The Swiss mathematician Jacob Steiner (1796–1863) posed the following problem: What value of *n* in the expression $\sqrt[n]{n}$ gives a maximum value? (The expression $\sqrt[n]{n}$ means the *n*th root of *n*.) If we let *n* equal, say, 1.99, the identity equals 1.41311473 . . . . If we let *n* equal 2, the identity equals the square root of 2, which is 1.4142135 . . . , which is slightly larger than the previous result. If n equals 2.5, the identity equals 1.44269990 . . . , which is larger again. If n equals 3, the identity equals 1.44224957 . . . , which is slightly smaller than the previous result. It turns out that the value of $\sqrt[n]{n}$ is at a maximum when n equals e. In other words, the e root of e gives a maximum result ($\sqrt[e]{e}$ equals 1.44466786 . . .).[23]

The number e is also intimately involved in the solution to the following problem: At what value of $1/x$ is the expression $1/x^{1/x}$ at a minimum value? The solution is when the value of $1/x$ equals $1/e$. In other words, $1/e^{1/e}$ equals 0.69220062 . . . , which is the minimum value of the expression $1/x^{1/x}$.[24]

The numbers e and π appear in Stirling's approximation (sometimes known as *Stirling's formula*), which may be used to *approximate* a factorial.[25] (One factorial = 1; two factorial = 2 · 1 = 2; three factorial = 3 · 2 · 1 = 6, and so on.) As I've mentioned already, the factorial is usually written with an exclamation sign after the number. For example 3! = 3 · 2 · 1. Stirling's approximation is:

$$n! \approx \sqrt{2\pi n} \cdot \left(\frac{n}{e}\right)^n.$$

If you let n equal 5, for example, Stirling's approximation gives 118.019165. . . . The correct answer is that five factorial (5!) equals 5 · 4 · 3 · 2 · 1, which equals 120. As n gets infinitely large, the percentage error between the approximate and true result for $n!$ decreases.

As you are well aware, Owen, the number e crops up in unexpected places in mathematics. For example, the solution to the following problem,[26] as the value of n approaches infinity, is e:

$$\left(\frac{2n+1}{2n-1}\right)^n.$$

(If a scientific calculator is nearby, punch in, say, 50, for the value of *n in the above formula*. The formula then yields 2.71837 . . . . As the value of n increases toward infinity, the formula approaches the value of e, or 2.718281828 . . . .)

Incidentally, I should mention that the number e is the unique positive number[27] for which the following holds:

$$\frac{e^x - 1}{x} = 1, \text{ as the value of } x \text{ approaches zero.}$$

The following property of e is also notable:

$$e \text{ is the limit of } (1+x)^{\frac{1}{x}} \text{ as the value of } x \text{ approaches zero.}$$ [28]

The number e is also the limit of the following identity:

$$n/^{n}\sqrt{n!} = e \text{ as } n \text{ approaches infinity.}$$

For example, if n equals 5, the identity equals 1.91925 . . . . When n equals 10, the identity gives 2.20812 . . . . When n equals 20, the identity gives 2.40837 . . . . When n equals 40, the identity gives 2.53669 . . . . When n equals 50, the identity gives 2.56630 . . . . When n equals 60, the identity gives 2.58711 . . . . In other words, as the value of n increases toward infinity, the identity gives a value that approaches the value of e.

As you pointed out in your chapter, many beautiful formulae link the two great transcendentals, e, π, with i, the number that is the square root of negative 1. These are three of the five fundamental identities in mathematics, the other two being 0 and 1. A very famous formula (probably the most famous of all) links all five identities together. The formula is $e^{i\pi} + 1 = 0$.[29] The equation, known as *Euler's identity*, contains three fundamental mathematical operations: addition, multiplication, and exponentiation (all appearing, with e, i, π, and 1 on one side of the equality sign. On the other side of the equality sign there is . . . nothing.) Over the decades, this formula has been described by many of the world's greatest mathematicians as being probably the most beautiful in the whole of mathematics.[30]

Here are two curiosities involving answers that are close to the value of the number e.

The letter e is the fifth letter of the English alphabet. When I was a child of nine, I discovered the following little approximation of e, which involves the integers from 1 to 5:

$\sqrt{1} \cdot \sqrt{2} + \sqrt{3} + \sqrt{4} + \sqrt{5}$ equals 7.382332347. This is close to the value of e^{2}. In other words,

$$e \approx \sqrt{1} \cdot \sqrt{2} + \sqrt{3} + \sqrt{4} + \sqrt{5} \approx \sqrt{7.382332347\ldots} \approx 2.7170447\ldots$$

When I was just ten years old, I realized that when 10^2 is divided by the value of e, the answer is 36.7879441 . . . . This result is very close to 37. So just for fun I decided to obtain the value of 3^7. This equals 2,187, which, by the way, equals $(2 + 1^8)^7$. I transferred the end digit of the number 2,187 to the second position in the number and obtained 2,718, which are the first four digits of the number *e*.

Beginning at the 99th decimal digit of *e*, the following ten digits appear: 7427466391. Concatenate these digits to form the number 7,427,466,391. This is the first ten-digit prime number that is found in the consecutive digits of *e*.

Finally, here are two little puzzles for your readers.

(1) Suppose you put 1,000 dollars in a bank account that earns you 5 percent compound interest per annum. How long will it take for your initial investment of 1,000 dollars to double?

Here's an easy way to calculate approximately how long it will take for your money to double. It's an interesting shortcut that your readers might be interested in knowing. Divide 70 by the annual interest rate. In this case, the annual interest rate is 5 percent. Therefore, divide 70 by 5. The answer is 14. Therefore, money placed at 5 percent compound interest will double in *approximately* 14 years.

Your readers might enjoy finding out why the shortcut works and why the shortcut involves the number 70. The number *e* is intimately involved in the solution.

(2) An old problem asks us to determine, without actually calculating the two values, or without using calculus, which of these two expressions is the greater: e^π or π^e. There are a number of approaches one may take to solve the problem. I found the following delightful and simple solution on www.math.stackexchange.com, which was posted on November 8, 2010.

Talk to you soon.
Dr. Cong

SOLUTIONS

(1) Dr. Cong's shortcut involved 1,000 dollars being placed in a bank account earning 5 percent compound interest per annum. The good

doctor said that 70 divided by the rate of interest gives the approximate number of years it takes for the money to double. You might wonder where the 70 comes from.

From the question we can write the following equations:

$$2 \cdot 1000 = 1000\,(1.05)^x.$$

Divide across the equation by 1,000 to simplify it:
This yields

$$2 = 1.05^x.$$

Thus 2 equals 1.05 raised to the power of x.
Thus

$$\ln 2 = \ln 1.05\, x$$

or

$$\ln 2 / \ln 1.05 = x$$

or

$$0.6931471 / 0.048790164 = 14.20669908\ldots.$$

Therefore, money placed at 5 percent compound interest per year will double in approximately 14 years.

Here's a brief explanation of the above solution: The ln of 2 equals 0.6931471 . . . because $e^{0.6931471} = 2$. The ln of 1.05 is 0.04879016 . . . because $e^{0.0487901} = 1.05$. (The ln of $1 + r$ will be close to r.) Dividing 0.69314718 . . . . by 0.04879016 . . . equals 14.2067003 . . . . Thus, money earning 5 percent compound interest per annum will double in 14.2067003 . . . years. (In other words, $1.05^{14.2067003\ldots}$ equals 2.)

From this you can see how Dr. Cong's shortcut is derived and why the number 70 enters into the solution. Instead of dividing 0.69314718 . . . . by 0.04879016 . . . , simply divide 100 times 0.693170 by 100 times 0.04879016. This is approximately 70 divided by 5. The result gives the *approximate* number of years that it takes money to double while earning 5 percent compound interest per annum. Or the result gives the approximate number of years for a population to double if the population is increasing at the rate of 5 percent per year. (If a population is increasing at the rate of 10 percent per year, divide 70 by 10 to obtain 7. Thus the population will double in approximately 7 years.)

(1) You are asked to determine, without actually calculating the two values, which of the two expressions is the greater: e^{π} or π^{e}. The problem can be solved without calculus if you know the following infinite series for e^x: [31]

$$e^x = 1 + x + x^2/2! + x^3/3! + \ldots.$$

The first two terms are 1 and x. This sum of the first two terms is clearly *less* than the complete sum of e^x.

Thus for positive values of x one finds that e^x is clearly greater than $(1 + x)$.

Let $x = \pi/e - 1$.

Then $e^{\pi/e-1}$ is greater than $1 + \pi/e - 1$.

Or $e^{\pi/e-1}$ is greater than π/e.

We note that $e^{-1} = (1/e)$. Therefore, we can write: $(1/e)\cdot e^{\pi/e}$ is greater than π/e.

This inequality can be rewritten as:

$$\frac{e^{\frac{\pi}{e}}}{e} \text{ is greater than } \frac{\pi}{e}.$$

Multiply across the inequality by e to obtain:

$$e^{\frac{\pi}{e}} \text{ is greater than } \pi.$$

Raise both sides of the inequality to the power of e to obtain:

$$e^{\pi} \text{ is greater than } \pi^{e}.$$

This result is usually written in mathematical text as:

$$e^{\pi} > \pi^{e}.$$

(Here the > sign signifies "greater than.")

The value of e^{π} is 23.140692632 ... and the value of π^{e} is 22.459157771. . . .

CHAPTER 14

PASCAL'S TRIANGLE

The seventeenth-century mathematician and philosopher Blaise Pascal (1623–1662) first wrote about a curious array of numbers arranged in a triangular formation in 1653. That is why the name *Pascal's triangle* is used today to describe this number pattern. However, the triangle was known long before Pascal wrote about it. It is known, for example, that the triangular array appears in a Chinese book dated to 1303. Scholars today believe that Chinese and probably Indian mathematicians also knew about the triangular array long before that.[1]

The triangular array of numbers is simple to write down. The first number in the triangle is 1, which is written at the triangle's apex. The top row of Pascal's triangle is usually denoted Row 0. That top row contains the number 1. Imagine an invisible zero at each end of Row 0. The 1 is added to each zero and the two answers are written down underneath in Row 1. This gives:

Row 0 1
Row 1 1 1

Each of the two 1s in Row 1 is added to the zero at each end of that row, and also to each other, to give Row 2. Thus the first three rows look like this:

Row 0 1
Row 1 1 1
Row 2 1 2 1

The procedure is continued so that each number in each row is the sum of the two numbers immediately above it, assuming that there is an invisible zero at both ends of each row.

Here are the first ten rows of Pascal's triangle:

Row 0											1										
Row 1										1		1									
Row 2									1		2		1								
Row 3								1		3		3		1							
Row 4							1		4		6		4		1						
Row 5						1		5		10		10		5		1					
Row 6					1		6		15		20		15		6		1				
Row 7				1		7		21		35		35		21		7		1			
Row 8			1		8		28		56		70		56		28		8		1		
Row 9		1		9		36		84		126		126		84		36		9		1	
Row 10	1		10		45		120		210		252		210		120		45		10		1

Figure 14.1

In a similar way to the rows, the diagonals in Pascal's triangle are numbered Diagonal 0, Diagonal 1, Diagonal 2, and so on. The number of rows in the triangle is infinite and symmetrical. If we were to draw a line straight down through the middle of the triangle, we would find the entries on the right-hand side are identical to those on the left-hand side.

If we inspect the triangle, one of the first things we will likely notice is that the sum of the nth Row is 2^n. Thus the sum of the numbers in Row 0 is 2^0, or 1. The sum of the numbers in Row 1 is 2^1, or 2; the sum of the numbers in Row 2 is 2^2, or 4; the sum of the numbers in Row 3 is 2^3, or 8. And so on.

The sum of the numbers above Row n is $2^n - 1$. Thus the sum of the numbers above Row 2, for instance, is $2^2 - 1$, or 3. The sum of the numbers above Row 3 is $2^3 - 1$, or 7. And so on.

Consider the diagonals on either side of the triangle. The first diagonal, containing all the 1s, is denoted as Diagonal 0. The next diagonal, containing all the natural numbers, 1, 2, 3, 4, 5 . . . is designated as Diagonal 1. The next diagonal, designated Diagonal 2, contains the numbers 1, 3, 6, 10, 15 . . . . These are the famous triangular numbers, which we discussed in chapter 11. Every second triangular number is known as a hexagonal number. The formula for the nth hexagonal number is $(2n^2 - n)$. The series of hexagonal numbers begins 0, 1, 6, 15, 28, 45, 66, 91, 120 . . . . Every perfect number is a hexagonal number. In 2001, a British statistician named Henry Bottomley (1963–) who likes numbers and patterns, discovered that the hexagonal numbers are the number of divisors of numbers of the form 12^{n-1}. For example, the number of divisors of 12 is 6. (The divisors of 12 are 1, 2, 3, 4, 6, and 12, as each of these numbers divide evenly into 12.) The number of divisors of 12^2 is 15. (The divisors of 12^2, or 144, are 1,

2, 3, 4, 6, 8, 9, 12, 16, 18, 24, 36, 48, 72, and 144.) The number of divisors of 12^3 is 28; the number of divisors of 12^4 is 45, and so on.

The next diagonal is designated Diagonal 3. The numbers that appear in Diagonal 3 are 1, 4, 10, 20, 35 . . . . These are the tetrahedral numbers, or the *pyramidal numbers*, as they are sometimes called. What are tetrahedral numbers? Think of billiard balls being arranged in a pyramid pile. Each layer of the pyramid contains a triangular number of billiard balls, beginning with 1 at the top. Thus 1 is termed a tetrahedral number. The second layer contains 3 billiard balls; therefore, the sum of 1 and 3 is also termed a tetrahedral number. If you care to check, you will find that a five-layer pyramid contains 1 + 3 + 6 + 10 + 15, or 35 billiard balls. Consequently, 35 is a tetrahedral number.

The series for tetrahedral numbers is 1, 4, 10, 20, 35, 56, 84 . . . . The formula for the nth tetrahedral number is $\frac{n(n+1)(n+2)}{6}$. If we let n equal 1, the formula is $(1(1 + 1) \cdot (1 + 2)/6)$. This equals $(1 \cdot 2 \cdot 3)/6 = 1$. Thus the first tetrahedral number is 1. If we let n equal 2, the formula gives $(2(2 + 1) \cdot (2 + 2)/6)$. This equals $(2 \cdot 3 \cdot 4)/6 = 4$. Thus the second tetrahedral number is 4. If we let n equal 3, the formula gives $(3(3 + 1) \cdot (3 + 2)/6)$. This equals $(3 \cdot 4 \cdot 5)/6 = 10$. Thus the third tetrahedral number is 10. And so on.

The tetrahedral numbers crop up in surprising places. Suppose we write the following columns of integers so that the first column on the left contains the series of triangular numbers; the second column from the left contains the series of natural numbers; the next column from the left contains the multiples of 2; then the next column contains the multiples of 3, and so on. We will find that the sums of each of the successive rows are the tetrahedral numbers in order (see figure 14.2).

1,									
3,	1,								
6,	2,	2,							
10,	3,	4,	3,						
15,	4,	6,	6,	4,					
21,	5,	8,	9,	8,	5,				
28,	6,	10,	12,	12,	10,	6,			
36,	7,	12,	15,	16,	15,	12,	7,		
45,	8,	14,	18,	20,	20,	18,	14,	8,	
55,	9,	16,	21,	24,	25,	24,	21,	16,	9,

Figure 14.2

Consider momentarily the natural numbers in Diagonal 1 (figure 14.1). Suppose we want to sum the first eight natural numbers. To do so, look at the number directly below and to the right of the last number to be summed. In this case, that number is 36. So 36 is the sum of the first eight natural numbers. A similar computation can be done with the numbers in each of the diagonals. For example, to find the sum of the first five tetrahedral numbers, go down the list of tetrahedrals. We find 1, 4, 10, 20, and 35. The sum of these five tetrahedrals is 70, which is just below and to the right of 35, the last number to be summed.

The prime numbers have an intimate relationship with Pascal's triangle as well. To see this, consider a row whose row number is a non-prime. Say, Row 4. The numbers that appear in Row 4 are 1, 4, 6, 4, 1. Ignore each of the 1s at either end of the row. Notice that not all numbers in Row 4 are evenly divisible by 4. That is, the number 6 is not evenly divisible by 4. Now consider a row whose row number is a prime. Say, Row 5. Once again, ignore the 1s in the row. Notice that all other numbers in Row 5 are evenly divisible by 5. This is always the case. In Row n, if n is a prime number, all of its numbers (ignoring the 1s at either end of the row) are divisible by n.

Pascal's triangle contains some charming and interesting properties that someone new to the study of mathematics can understand. For example, consider Row 4. Sum the squares of all the numbers in Row 4. The answer, 70, is the middle number of Row 8. This property holds throughout the triangle. The sum of the squares of the numbers in Row n equals the central number in Row $2n$.

The triangle can be used as a ready reckoner to solve problems involving combinations. For instance, suppose we go to a fancy store, wishing to buy three shirts. Let's suppose that there are eight different-colored shirts in the store that are of a suitable size to choose from. How many selections can we make? To find the answer, go to the intersection of Diagonal 3 and Row 8. The number there is 56. Therefore, there are 56 selections we can make. If we mistakenly search for the intersection of Diagonal 8 and Row 3, we will find that that intersection does not exist in Pascal's triangle. So there is no alternative but to go to the right intersection: Diagonal 3 and Row 8.

This property of Pascal's triangle can be utilized in numerous ways. Let's consider a few examples. Suppose we're told that a couple have six children in their family. What is the probability that the couple have three boys and three girls?

To obtain the answer, go to the intersection of Diagonal 3 and Row 6. (Remember the first row is Row 0 and the first diagonal is Diagonal 0.) The number there is 20. The sum of the numbers in Row 6 is 2^6, or 64. Therefore, the probability that there will be three boys and three girls in a family of six children is 20/64 or 5/16. This equals 0.3125, or 31.25 percent.

What is the probability that in a family of seven children there will 2 boys and 5 girls? Go to the intersection of Diagonal 2 and Row 7. The number there is 21. The sum of the numbers in Row 7 is 2^7, or 128. Therefore, the probability that a family containing seven children consists of 2 boys and 5 girls is 21/128, or 0.1640625 (16.40625 percent).

What is the probability that if you toss nine coins, 4 heads and 5 tails will show up? Go to the intersection of Diagonal 4 and Row 9. The number there is 126. The sum of the numbers in Row 9 is 2^9, or 512. Therefore, the probability is 126/512, or 63/256. This equals 0.24609375, or 24.609375 percent. If instead of going to Diagonal 4 and Row 9 we went instead to Diagonal 5 and Row 9, we would still find the number at that intersection to be 126. Thus the probability would still equal 126/512. This is because the probability of choosing 4 heads and 5 tails from nine tosses must be the same as choosing 5 heads and 4 tails from nine tosses.

Suppose someone is selling tickets and she has just 11 left to sell. You decide to buy 4 of them. How many different selections can you make? Go to the intersection of Diagonal 4 and Row 11. The first eight terms of the 11th row of the triangle, reading from the left are 1, 11, 55, 165, 330, 462, 462, 330. The number at the intersection of Diagonal 4 and Row 11 is 330. Therefore, there are 330 different selections you can make.

The above properties of Pascal's triangle can be clearly seen if we write the triangle as shown in figure 14.3.

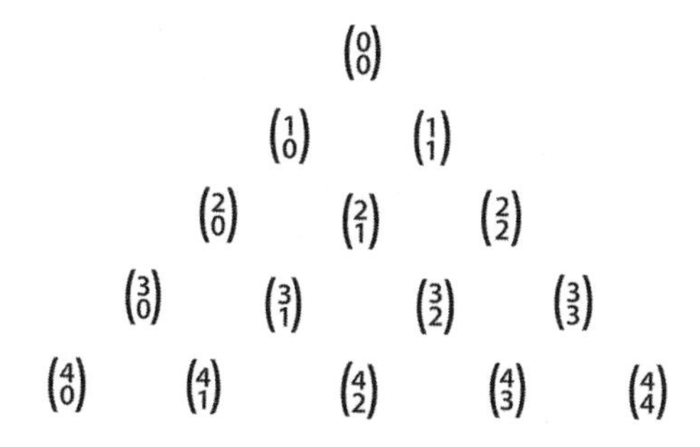

Figure 14.3

Consider any row in the triangle in figure 14.3. Say Row 3. The first entry tells us that the number that is in this position in Pascal's triangle shows us how many ways zero objects can be chosen from 3 objects. There is only 1 way to do this: Don't choose any objects at all! The second entry in Row 3 tells us that the number that appears in this position in Pascal's triangle shows us how many ways there are to choose 1 object from 3 objects. There are 3 ways to do this.

The third entry that appears in this position in Pascal's triangle tells us how many ways 2 objects can be chosen from 3 objects. There are 3 ways of doing this. Finally, the last entry in Row 3 in figure 14.3 tells us that the number that appears in this position in Pascal's triangle shows how many ways there are of choosing 3 objects from 3 objects. There is only one way of doing this. So the numbers that appear in Row 3 of the triangle are 1, 3, 3, 1.

Gamblers are notorious for being suspicious. Many gamblers believe, for example, that the more often a fair coin is tossed, the more likely it is that the number of heads will *exactly* equal the number of tails. Pascal's triangle illustrates that this belief is a fallacy. First of all, the number of tosses must be even if the number of heads and tails are to be equal. Consider Row 4 of the triangle. This tells us that if we toss a coin four times, there are 2^4, or 16 possible outcomes, of which 6 will contain 2 heads and 2 tails. So with four tosses, the odds of an equal number of heads and tails are 6/16 or 0.375, or 37.5 percent.

Consider Row 6. This row tells us that if we toss a coin six times, there are 2^6 or 64 possible outcomes, of which 20 have an equal amount of heads and tails. Therefore, the odds of an equal amount of heads and tails with six tosses are 20/64 or 0.3125, or 31.25 percent.

Consider Row 8. This row tells us that there are 2^8 or 256 possible outcomes to eight tosses of a coin, of which 70 have an equal amount of heads and tails. Therefore, the odds of tossing eight coins and having an equal amount of heads and tails are 70/256 or 0.2734375, or 27.34375 percent. Similarly, if we toss ten coins, the odds of having an equal amount of heads and tails are 252/1024 or 0.24609375, or 24.609375 percent. And so on. The odds of obtaining an exact equal amount of head and tails are declining as more and more tosses are made.

The triangle has many other uses in mathematics. For example, suppose we want to find the coefficients in the solution to the following algebra problem: Raise $(x + y)$ to the power of 2. Instead of actually doing the multiplication, we can use Pascal's triangle to find the coefficients in the answer.

This is easily obtained. We are asked to solve $(x + y)^2$. Therefore, look at Row 2 of the triangle (recall that the top row of the triangle is Row 0) to find the relevant coefficients. The numbers in Row 2 of Pascal's triangle are 1, 2, 1. The numbers 1, 2, 1 will be the coefficients in the answer. (We see that the power number determines the row chosen.)Therefore, the solution to $(x + y)^2$ is $1x^2 + 2xy + 1y^2$. Note that as we go from left to right in that solution, the exponents of x decrease and the exponents of y increase. This is always the case.

In general the coefficients of the expansion of $(x + y)^n$ are found in the nth row of the triangle. For instance, the solution of $(x + y)^3$ is $1x^3 + 3x^2y + 3xy^2 + 1y^3$. These coefficients, 1, 3, 3, 1, are found in the third row of the triangle. (It is customary, incidentally, for mathematicians to omit the coefficient of 1 when writing algebra equations.) Thus the solution to $(x + y)^2$ is usually written as $x^2 + 2xy + y^2$; the solution to $(x + y)^3$ is usually written as $x^3 + 3x^2y + 3xy^2 + y^3$. And so on.

Here is a beautiful property of the triangle that is not so well known. Let's consider any row—say, Row 4 (see figure 14.4). Consider the number one entry in Row 4, which is 4 (recall that the zero entry is 1). Consider the number two entry in Row 6, which is 15; and consider the number three entry in Row 5, which is 10. Obtain the product of these three numbers: $4 \cdot 15 \cdot 10$. The answer is 600. Consider now the first entry of Row 5, which is 5; the second entry of Row 4, which is 6; and the third entry of Row 6, which is 20. Obtain the product of these three numbers. The answer is also 600. This property holds throughout the triangle. Generally, this identity is written in mathematical language as $(n - 1C1) \cdot (nC3) \cdot (n + 1C2) = (nC1) \cdot (n - 1C2) \cdot (n + 1C3)$.

Here n stands for the row number, which in this case is Row 5. Thus $(n - 1C1)$ means the $(5 - 1)$ row, or Row 4, and the $C1$ means the number of combinations or the number of ways of selecting 1 object from that row number. In other words, $(n - 1C1)$ means the 4th row and the number of ways of selecting 1 object from 4, which equals 4. The $(nC3)$ means the 5th row and the number of ways of choosing 3 objects from 5, which equals 10. The $(n + 1C2)$ means the 6th row and the number of ways of choosing 2 objects from 6, which equals 15. Thus $(n - 1C1) \cdot (nC3) \cdot (n + 1C2)$ means $4 \cdot 10 \cdot 15$. This equals 600. In a similar way we find that $(nC1) \cdot (n - 1C2) \cdot (n + 1C3)$ equals $5 \cdot 6 \cdot 20$, which also equals 600.

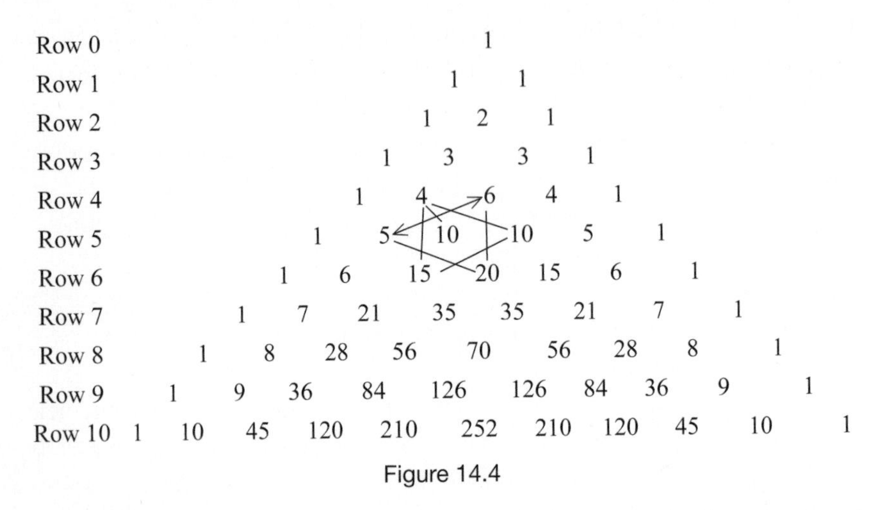

Figure 14.4

This beautiful property and many other similar properties have intrigued mathematicians over the years. As a result, mathematicians became convinced that Pascal's triangle was full of many other wonders just waiting to be discovered. This belief resulted in many professional and amateur mathematicians searching for and discovering many other number curiosities in the triangle.

Down through the decades of the twentieth century, mathematicians wondered if the famous number π could be found in Pascal's triangle. In 2007, an amateur mathematician from Colombia named Jonas Castillo Toloza found that the triangular numbers that are found in Diagonal 2 of Pascal's triangle can be used to obtain the value of π. Toloza discovered the following equation[2]:

$$\pi = 2 + \frac{1}{1} + \frac{1}{3} - \frac{1}{6} - \frac{1}{10} + \frac{1}{15} + \frac{1}{21} - \frac{1}{28} - \frac{1}{36} + \ . \ . \ . \ .$$

Daniel Hardisky, a retired civil engineer from Baltimore, Maryland, also discovered π hidden in Pascal's triangle. His unearthing of π involves the tetrahedral numbers, which are found in Diagonal 3 of the triangle (recall that the first diagonal is called Diagonal 0).[3] More precisely, Hardisky's equation involving π includes every second tetrahedral number as a denominator, beginning with 4. The beautiful equation Hardisky discovered is:

$$\pi = 3 + \frac{2}{3} \cdot \left(\frac{1}{4} - \frac{1}{20} + \frac{1}{56} - \frac{1}{120} + \frac{1}{220} - \ . \ . \ . \right)$$

The famous number *e*, which equals 2.718281828459 . . . , was also found in Pascal's triangle in 2009 by an inventor and mathematician in New Haven, Connecticut, Harlan J. Brothers. To find *e* calculate the products of each row in Pascal's Triangle:

Row 0	1
Row 1	1 1
Row 2	1 2 1
Row 3	1 3 3 1
Row 4	1 4 6 4 1
Row 5	1 5 10 10 5 1
Row 6	1 6 15 20 15 6 1
Row 7	1 7 21 35 35 21 7 1
Row 8	1 8 28 56 70 56 28 8 1

Figure 14.5

Consider the triangle shown in figure 14.5. The products of the first eight rows, beginning at row 0, are: 1; 1; 2; 9; 96; 2,500; 162,000; 26,471,025; 11,014,635,520. Let the products of the numbers in the *n*th row equal *Pn*. Then *e* is the limit as the value of *n* approaches infinity in the following equation:

$$\frac{\left(\left(Pn-1\right)\left(Pn+1\right)\right)}{\left(Pn\right)^{2}}.$$

Let's take two examples of how the equation is used to come closer and closer to the true value of *e*. Let *n* equal 4. The product of the numbers in Row 3 is 9, and the product of the numbers in row 5 is 2,500. Multiply these two numbers together, obtaining 22,500. Divide the result by the square of the product of the numbers in Row 4, which is 96^2. The answer is 2.4414062 . . . . This is about 89.8 percent of the true value of *e*.

If we substitute 5 for *n*, we would find that the product of the numbers in Row 4 is 96 and the product of the numbers in Row 6 is 162,000. Multiply these two numbers together to obtain 15,552,000. Divide this result by the square of the product of the numbers in row 5, which is $2{,}500^2$. The answer is 2.48832. This is about 91.5 percent of the true value of *e*. This result is a little closer to the value of e than the previous result. As we perform this procedure on the descending rows of Pascal's triangle, the answer gets closer and closer to the true value of e.

Mathematicians, scientists, philosophers, and mystics have been known to express awe and reverence when they contemplate the astonishing fact that two of the most ubiquitous numbers in the whole of mathematics, π and e, can be found in Pascal's triangle.

The study of pure mathematics is defined as the study of mathematics in order to discover the inherent beauty in the subject, rather than asking where mathematics can be applied in various sciences such as engineering, physics, or navigation. For those who may ask if there is some practical use of Pascal's triangle in the physical world, it is incredible to learn that the coefficients in the triangle are associated with magnetic resonance imaging, often referred to as MRI scanning.

The sum of each row of numbers in Pascal's triangle equals a power of 2. For example, the number at the top of the triangle is 1, which equals 2^0. Row 1 contains two 1s, which sum to 2, which equals 2^1. Row 2 contains 1, 2, and 1. Those numbers sum to 4, which equals 2^2. And so on.

You may be interested to learn why $2^0 = 1$. When we divide powers of a similar positive number, a shortcut exists in which we subtract the smaller power from the other to get the answer. For example, if we divide 2^9 (512) by 2^5 (32), we subtract 5 from 9, obtaining 4. Thus the answer is 2^4, or 16. Similarly we find that, say, 7^5 (16807) divided by 7^3 (343) equals 7^2, or 49. And so on.

What happens if we divide, say, 2^9 by 2^9. We know that 512 divided by 512 equals 1. The shortcut tells us that 2^9 divided by 2^9 equals $2^{9-9} = 2^0$. Since 512 divided by 512 equals 1, it follows that 2^0 equals 1. Or 7^5 divided by 7^5 equals 7^{5-5}. This equals 7^0, which is also equal to 1. Thus any positive number raised to the power of zero equals 1.

Those mathematicians who argue that we invent mathematics will tell you that any positive number raised to the power of zero is equal to 1, but that that result is purely as a consequence of convention by mathematicians around the world.

In other words, the mathematicians who argue that mathematics is simply an invention of the human mind infer by their position that mathematical results are a consequence of rules made by human minds, and could—if humans deemed it so—be different from what they are.

I find this argument difficult to accept. I have many reasons for my belief in the existence of a mathematical structure outside of human minds. The following is one simple argument. If extraterrestrial intelligent life is ever discovered, and we can communicate with them, I cannot believe that they would argue that the number 23 in the decimal system (as we know it) is not prime. The fact is that 23 is prime in the decimal system not because human beings have decided that it is, but because it *is*.

Suppose we asked one of these aliens to take 23 pebbles and arrange all of them in a rectangular number of rows, so that there is more than one row and each row contains the same number of pebbles. The alien could not do it. The best he could do is line the 23 pebbles up in one straight line. He could not arrange them in more than one row with the same number of pebbles in each row. Why? Because the only divisors of 23 are 1 and 23. The number 23 is prime. Mathematical reality tells us it's prime, and we have no option but to accept that fact.

My argument is basically this: We have to accept mathematical reality as it is. It imposes its truths upon us. I believe that Pascal's triangle illustrates in one simple way that mathematical truths exist "out there" in an abstract way in the external world, and not—as some mathematicians and philosophers would have us believe—inside human minds only. Here's why I believe this.

We know that Pascal's triangle contains an infinite number of rows where the sum of each row is a power of 2 beginning at Row 0, which has just one number, the number 1, in it. The very top row, Row 0, has a sum equal to 1. Pascal's triangle shows us that 1 is equal to 2^0. It seems strange that Pascal's triangle "knew" this mathematical fact many centuries ago, long before mathematicians around the world decided by "convention" that any number raised to the power of zero equals 1.

Pascal's triangle may give a hint to the answer of one of the deepest questions in philosophy. Down through the centuries, philosophers from Leibniz to Schopenhauer to Heidegger have wondered: Why does the universe exist at all? Why is there something rather than nothing?

The question is unlike any other. Some have even said that the question is meaningless. It is probably fair to say that most people have never encountered the question. Those who have often dismiss it, thinking it is not worth worrying about. It is likely that most people go through their entire lives without ever considering this question.

When you think about the question (which has been referred to as the super-ultimate question in philosophy), perhaps the first thought that comes to mind is that things would be much simpler if nothing, absolutely *nothing* at all, existed for all time. Since nature usually takes the more efficient and economical solution to problems, we can ask: Why did nature not "choose" to have the apparently simplest scenario in place from all eternity—that is, that absolutely nothing at all should exist for all eternity?

The super-ultimate question isn't answered by saying that a supernatural being or god is responsible for the existence of all things. That answer only transfers the question from one peg to another. One can immediately ask: Why does God exist?

Many of those who understand the significance of the super-ultimate question have said that the question reemerges in their mind every few months. The question strikes them when they least expect it. These people have reported feelings of fear, awe, amazement, and astonishment, as they contemplate the fact that the universe exists. These feelings are often accompanied with sincere humility as they ponder the super-ultimate question: Why does the universe go to the trouble of existing? Why *something*, rather than *nothing*? Would it not be much simpler, much less troublesome, if *nothing* at all existed? What law of nature should determine that things exist?

The American philosopher William James (1842–1910) described the super-ultimate question as being the "darkest in all of philosophy."[4] Many philosophers argue that the question is meaningless, because any explanation that is put forward for the world's existence, be it a cosmic egg or divine creator, is part of the something to be explained.

One view in modern science seems to suggest that the universe emerged from a quantum void about thirteen and a half billion years ago. The quantum void, many scientists believe, preexisted the universe. It probably exists for all time and contains quantum laws that somehow brought forth the existence of at least one universe. If this is true, then you can ask, where did the quantum laws come from? And why are those quantum laws framed the way they are and not some other way? It appears that there is no answer to the super-ultimate question.

Of course, you could choose to believe in an oscillating universe. Such a universe comes into existence, expands for billions of years and then due to gravity contracts and eventually collapses to a singularity. Then from the singularity another universe bursts forth and the cycle repeats endlessly. This could be the case for all eternity, with no beginning and no end.

If you believe in an oscillating universe, then you could ask, why is this the nature of things? Why are things this way and not another way? Once again the super-ultimate question seems to be staring us in the face.

If, however, you believe that mathematical reality has always existed, then that belief may offer an explanation of why something exists rather than nothing. It may be that the laws of probability determine that it is much more likely that *something* will exist rather than *nothing* at all!

What does Pascal's triangle have to do with all of this? Well, let's see. Consider the following fact. The sum of the numbers in Row n of Pascal's triangle is 2^n. Because of this probability property in Pascal's triangle, the following identity holds.

$$(nC0)+(nC1)+(nC2)+\ldots+(nC(n-1))+(nCn)=2^n.$$

Figure 14.6

Here, for example, $nC1$, means the number of ways 1 object can be chosen from n objects. (Thus C equals "the number of combinations.")

For example let n equal 5. The equation in figure 14.6 then becomes:

$$(5C0)+(5C1)+(5C2)+(5C3)+(5C4)+(5C5)=2^5.$$

This equals

$$(1)+(5)+(10)+(10)+(5)+(1)=2^5.$$

Figure 14.7

Let us explain this identity in everyday terms. Suppose a man in a fancy store has five shirts to select from. The above equation tells us that there is 1 way of selecting zero shirts; there are 5 ways of selecting 1 shirt; there are 10 ways of selecting two shirts; there are 10 ways of selecting three shirts; and there are 5 ways of selecting four shirts from the five. Finally there is just 1 way of selecting all five shirts.

The total number of selections that the man can make equals, $1 + 5 + 10 + 10 + 5 + 1 = 32 = 2^5$. Of these 32 selections, 1 selection (the first one) involves the man selecting *no* shirts.

Using mathematical language, we say that the total number of subsets of a set is 2^n, where n is the number of sets. Thus with five objects, such as shirts, there are a total of 2^5 or 32 subsets. Crucially, these 32 subsets include one set that contains nothing at all. This set is referred to by mathematicians as the *empty set*. This tells us that from a mathematical point of view the laws of mathematics take into account that the concept of *nothing* can exist.

Of course we may well conclude that it is extraordinary how Pascal's triangle reveals that the *empty set* exists. But how is all this connected to the super-ultimate question?

Let's see.

If *nothing* at all were to exist, then only one set (the *empty set*) exists. The total number of subsets of the empty set is 2^0, which equals 1. Thus 1 subset of the *empty set* exists, and that is the *empty set* itself. On the other hand, if 1 particle only were to exist, the total number of subsets would be 2^1 or 2. One of these subsets would contain 1 particle and the other set (the *empty set*) would contain nothing at all. In this scenario, the probability that *nothing* would exist would be 50 percent. That is the highest percent that *nothing* exists will ever attain.

If just 20 particles were to exist, the total number of subsets would be 2^{20}, or 1,048,576. That is over one million subsets. Only one of these subsets, however, would be the *empty set*.

We see from this that the *empty set* is extremely unlikely to be the only set in existence, simply because it has such a low probability of occurring. Of all possible sets, there can be only 1 *empty set*. In other words, there can be an infinite number of ways *something* can exist but only one way that absolutely *nothing* (the *empty set*) can exist. Therefore *something* exists because it is much more likely than *nothing*.

Thus it appears that if nature were to "choose" (whatever that may mean) whether there was to be *something* or *nothing*, it is vastly more likely that there will be *something*. It would appear that the scenario where *nothing* at all exists would have an extremely low probability of occurring. Consequently, because of the law of probability theory, it is overwhelmingly likely that *something* will exist rather than *nothing*.

Thus it is possible that the universe exists because of the laws of probability theory. Pascal's triangle illustrates many properties of probability theory, and perhaps these probability laws are one way of explaining why there is *something* rather than *nothing*.

Of course I should say that science is far better suited to answering *what*, *where*, and *how* questions rather than *why* questions. These *why* questions are only meaningful when there is a broader context that we can refer to in attempting to answer the question. For example, if I find that my car is unable to start, I can ask why. I eventually figure out that it does not start because there is no gas in its fuel tank. I then ask why there is no fuel in the tank. I eventually answer that this is so because I forgot to purchase gas yesterday. And a line of similar reasoning continues, inquiring why I forgot to purchase the gas. And so on.

Each time, I can answer the question by referring to some other thing or event *within the universe*. But when you ask a *why* question about the origin of the universe or why it exists, there is no outside context, no encompassing fabric, that we can refer to. The universe is, by definition, everything that exists. Therefore, we cannot say that this or that caused the universe to exist, unless we refer to something that exists outside of the universe.

In recent years, many prominent scientists have come to accept the view that a timeless, eternal, quantum void has always existed and always will, whose quantum laws are mathematical and whose laws nature mysteriously follows. These scientists believe that our universe, with all of its complexity and beauty, emerged from the quantum void about thirteen and a half billion years ago.

It may be that this event has only occurred once, or it may well have occurred more than once. It seems reasonable to me to believe that since this event occurred at least once, then it is possible, probably very likely, that it has happened an enormous number of times—perhaps an infinite number of times. Thus it may well be that a multiverse exists, in which our enormous universe is just one tiny grain of sand on a sandy beach that is infinitely long.

If—as science appears to reveal—the quantum laws of the void are mathematical, then it follows that mathematical reality has always existed.

The super-ultimate question then reduces to why the void—with some sort of mathematical structure—exists. It is believed that this mathematical void has always existed. It is eternal. There probably is no explanation why it exists. It just *is*.

We are once again facing the super-ultimate question head-on. The question appears to be impenetrable.

I will say this. If the quantum void exists and if it has a mathematical structure, perhaps in some strange way the laws of mathematics determine—because of the theory of probability—that the laws of mathematics exist.

I sent off a draft of this chapter to Dr. Cong to see if he had any constructive comments to make. As I expected, he was not shy about giving his opinion. Here is the e-mail he sent me:

Hi, Owen,

Thank you for sending the draft of your chapter on Pascal's triangle

Because of the combinational structure that exists in Pascal's triangle, the triangle can be used to solve problems like this:

Suppose you wish to back a horse in each of four races. Go down to Row 4 in Pascal's triangle. The numbers in that Row are 1, 4, 6, 4, and 1. The first 1 (let's say the number 1 on the far left) tells you that there is just 1 way of backing *none* of the four horses. The next number, 4, tells you that there are just four ways of backing *one* of the four horses. The next number, 6, tells you that there are just six ways of backing *two* of the four horses. The next number, 4, tells you that there are just four ways of backing *three* of the four horses. The last number, 1, tells you that there is just one way of backing *all* four horses.

The number of odd entries in row n of Pascal's Triangle is 2 raised to the number of 1s in the binary expansion of n. For example, suppose we want to know how many odd entries are in Row 6 of the triangle. Well, 6 written in binary is 110. So there are two 1s in the binary expression of the number 6. Therefore, there are 2^2, or four, odd numbers in Row 6 of the triangle. Row 6 of Pascal's triangle is 1, 6, 15, 20, 15, 6, 1. As you can see, there are four odd numbers in this sequence.

There are many other beautiful patterns in Pascal's triangle. Many of these were discovered down through the centuries. However, it was only in recent decades that the Fibonacci numbers were found in Pascal's triangle. Recall that the Fibonacci sequence is 1, 1, 2, 3, 5, 8, 13, 21, 34, 55, 89, 144, and so on, each number after the third being the sum of the preceding two numbers. The Fibonacci sequence emerges on the far right as the sum of numbers that have a slanted line running through them in Pascal's triangle as shown in figure 14.8:

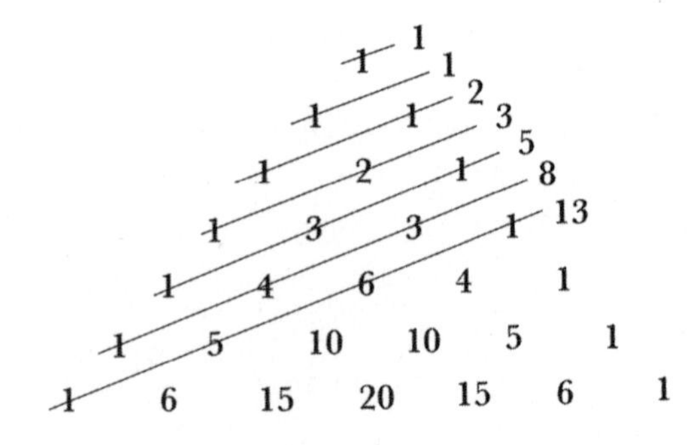

Figure 14.8

There is a simple way of finding Pythagorean triples in Pascal's triangle. Consider each odd-numbered row in the triangle, beginning with Row 3. Consider entry 0 and entry 1 in that row. They are 1 and 3. Sum the squares of both numbers and divide the result by 2, which gives you 5. Thus 5 is the hypotenuse of a right triangle. One less than 5, 4, is the longer leg. The row we chose is Row 3. Therefore, the shorter leg is 3. Thus the resulting right triangle has the sides 3, 4, and 5.

Suppose we begin with Row 5. Consider entry 0 and entry 1 in that row. They are 1 and 5. Sum the squares of both numbers and divide the result by 2 to obtain 13. Thus 13 is the hypotenuse of a right triangle. One less than 13, or 12, is the longer leg. The row that we chose is Row 5. Therefore, the shorter leg is 5. Thus the resulting right triangle has the sides 5, 12, and 13.

The fact that Pascal's triangle is connected to the Fibonacci sequence and the Pythagorean theorem fascinates mathematicians, scientists, philosophers, and interested laypersons. Such discoveries have led mathematicians to ask what other mathematical gems may be hidden within the triangle. One question that was asked over the centuries was if the triangle had any connection with perfect numbers.

All perfect numbers known to date are even. No odd perfect numbers are known to exist. (No one has proved, however, that odd perfect numbers cannot exist.) A perfect number is a number that is the sum of all its divisors (except itself). The smallest perfect number is 6. Its divisors are 1, 2, 3, and 6. The number 6 equals the sum of 1, 2, and 3. The first four perfect numbers are 6, 28, 496, and 8,128. Many mathematicians and number enthusiasts find it astonishing that *all* even perfect numbers appear in Pascal's triangle.

The formula for perfect numbers is

$$2^{p-1}\left(2^p - 1\right), \text{ where } p \text{ and } 2^p - 1 \text{ are both prime numbers.}$$

If we color all of the odd numbers that appear in Pascal's triangle in black and all the even numbers in gray, the image as shown in figure 14.9 emerges. The first gray triangle in Pascal's triangle in figure 14.9 contains 1 even number (the number 2). The second gray triangle contains 6 even numbers (4, 6, 4, 10, 10, 20). The third gray triangle (which is partially shown) has seven numbers to each side and therefore contains 28 even numbers. The fourth gray triangle (not shown here) contains 120 even numbers. The fifth gray triangle (not shown here) contains 496 even numbers. The numbers 6, 28, and 496 are the three smallest perfect numbers.

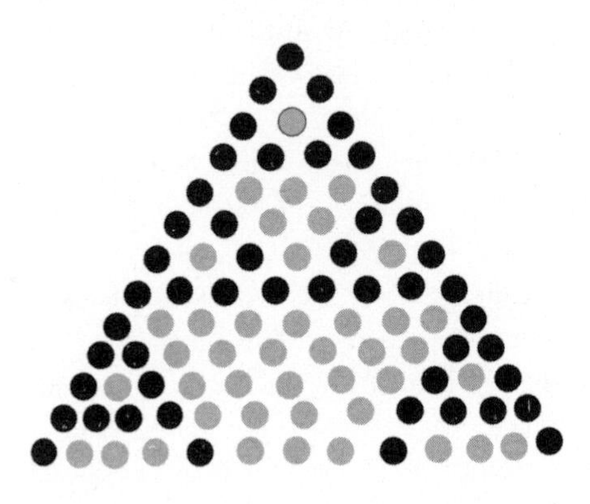

Figure 14.9

The number of dots in each gray triangle as one goes down through Pascal's triangle can be shown to be equal to $2^{p-1}(2^p - 1)$. This is, of course, the formula for perfect numbers, when $2^p - 1$ equals a prime number. Thus as we go down through Pascal's triangle, the number of numbers in each of the infinite number of gray triangles will be equal to $2^{p-1}(2^p - 1)$. When $2^p - 1$ is a prime, the number of numbers in the subsequent gray triangle will be a perfect number.

The expression $2^p - 1$ is a prime when p equals 2, 3, 5, 7, 13 . . . . As you mentioned in an earlier chapter, these primes are known as Mersenne primes, named after Marin Mersenne (1588–1648), a French mathematician and philosopher who studied them. It is not known if $2^p - 1$ *always* generates a prime number, but when it does, it also generates a perfect number. Thus *if* there are an infinite number of Mersenne primes, then it follows that there are an infinite number of perfect numbers.

However, it is not known if there are an infinite number of Mersenne primes. Most mathematicians believe that there probably are and, consequently, they also believe that there are an infinite number of even perfect numbers.[5] Although we cannot be certain that there are an infinite number of perfect numbers, we can say with certainty that every *even* perfect number appears—sooner or later— as the number of even numbers in the relevant gray triangles in Pascal's triangle.

Perfect numbers appear in Pascal's triangle in a second way. The number at the intersection of Diagonal 2 and Row 2^n of the triangle, when $2^n - 1$ is a prime number, is a perfect number! Thus the first perfect number, 6, is found at the intersection of Diagonal 2 and Row 4. The second perfect number, 28, is found at the intersection of Diagonal 2 and Row 8. The third perfect, 496, is found at the intersection of Diagonal 2 and Row 32. And so on.

You will notice from figure 14.9 that the number of gray dots increases as we go farther and farther down the infinite number of rows in Pascal's triangle. This tells us that the vast majority of numbers that appear in Pascal's triangle are even.

Let's pick number 1 in Row 0. Traveling in a southeast or southwest direction from the top 1 in Row 0, there is just 1 path to either 1 in Row 1. Traveling in a southeast or southwest direction from the top 1 in Row 0, there are just 2 paths we can travel to reach number 2 in Row 2 of the triangle. Traveling in a southeast or southwest direction from the top 1 in Row 0, there are just 3 paths to reach either one of the 3s in Row 3 of the triangle. And so on. (See figure 14.10.) Thus if we choose any number at random in Pascal's triangle, say, 20, we will find that there are exactly 20 paths from the top 1 in the triangle, if we travel in a southeast or southwest direction. If we choose 70, there are 70 paths leading to the number 70. And so on.

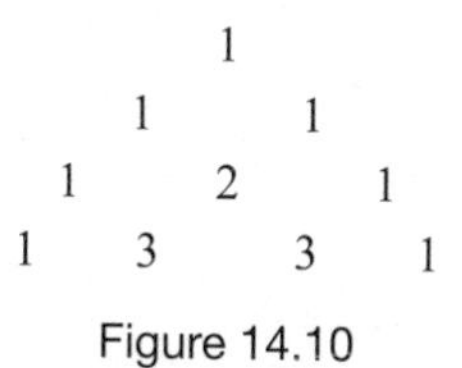

Figure 14.10

There are obviously an infinite number of 1s in Pascal's triangle. David Singmaster, a retired professor of mathematics from London, asked in 1971 how many times an integer (besides 1) occurs in Pascal's triangle. The number 2 appears just once; the number 3, twice; the number 6, three times; and the number 10, four times. The number 120 appears six times. The number 3,003 appears eight times in the triangle—it is the only number (besides1) that appears eight times. It is not known if any number appears more than eight times in Pascal's triangle.

Here is a little number play related to Pascal's triangle and to the life of Pascal:

Blaise Pascal was born in 1623 and died in 1662. The digits of 1662 sum to 15. Reverse the digits of 15 to obtain 51. The sum of the numbers from 1 to 51 is 1326. Transport the digit at the front of that number to the end to obtain 3261. The reverse of 3261 is 1623. That was the year Blaise Pascal was born.

Pascal's triangle can be built up by simply adding numbers together. An adding procedure also connects Pascal's birth year to the year of his death.

Let's look at his birth year of 1623.

$$1623 \rightarrow 1+6+2+3=12 \rightarrow 1623+12=1635$$
$$1635 \rightarrow 1+6+3+5=15 \rightarrow 1635+15=1650$$
$$1650 \rightarrow 1+6+5+0=12 \rightarrow 1650+12=1662$$

Blaise Pascal died in 1662. Isn't it interesting how much is tied to Pascal's triangle? Thank you again, Owen, for showing me your work on this fascinating topic.

Yours sincerely,

Dr. Cong

CHAPTER 15

SOME STRANGE AND REMARKABLE COINCIDENCES

Most people, I think it is fair to say, like coincidences. When an amazing coincidence is brought to our attention, the response is usually one of disbelief. Most people usually feel touched and bewildered by these apparently rare occurrences.

Why do coincidences occur? Two basic explanations are usually put forward. The first one argues that coincidences happen as a result of supernatural influences. But those of us who have a scientific way of looking at things recognize that coincidences occur because of the mathematical laws of probability. Just think about it. There are over seven billion people on this planet, doing hundreds of different things each day! Thus, every single day of the year, there are billions of events happening. From this perspective, it would be very unusual if surprising events did NOT occur every now and then.

Viewed from this perspective, the long arm of coincidence does not appear very long at all. However, when surprising coincidences do occur, most of us are pleasantly surprised and ask ourselves: What are the chances of that happening?

Here are nine short examples of coincidences occurring in the everyday life of people.

1. On May 25, 1979, author Judith Wax was on Flight 191 from O'Hare International Airport in Chicago to Los Angeles International Airport. She was on her way to Los Angeles to promote her recently published book, *Starting in the Middle*. The book was about middle-aged life. There were 258 passengers on board and 13 crew members. On page 191 of the book she was traveling to promote (note the page number and the flight number), Judith described her fear of flying in airplanes. She was one of

the 271 people killed when the plane crashed shortly after takeoff. In terms of casualties, it is to this day the worst airplane accident in US history.[1]

2. In the early 1970s, the well-known actor Anthony Hopkins was offered a major role in the movie based on the book *The Girl from Petrovoka* by George Feifer. To familiarize himself with the script, Hopkins traveled to London, England, and toured the bookstores in that city to locate a copy of the book, but to no avail. Later in the day, he made his way home, disappointed that he had not succeeded in finding a copy of Feifer's book. Hopkins went to Leicester Square Underground Tube station, where he waited for an underground train. While waiting, he noticed a book lying on a bench. Hopkins was astonished to find that the book was *The Girl form Petrovoka.*

   But the story gets even stranger. Two years later, Hopkins was in Vienna for the filming of the movie. The author of the book arrived on the set and started talking to him. In the course of conversation, Feifer told Hopkins that he (Feifer) did not have a copy of his own book. He said he had loaned his last copy of the book, containing his own annotations, to a friend who had subsequently lost the book in London. Hopkins produced the copy of the book he had found in Leicester Square Tube station. "Is this your book?" asked Hopkins, as he handed it to Feifer. Incredibly it was![2]

3. Jami Johnson absentmindedly left her wallet on the counter of a store in Clarkston, near Lewiston, Idaho, on Tuesday, December 11, 2007. Surveillance video footage later captured a man taking the wallet and walking out of the store with it.

   Two days later, on Thursday, December 13, the *Lewiston Tribune* had two separate photographs of a man on its front page. One photo was a still from the video camera reportedly showing a man on the previous Tuesday taking Johnson's wallet in the store and placing it in his coat pocket in Clarkston.

   Above that photo was another photo of a man decorating a shop-front window for Christmas. That second photograph had been taken by one of the *Lewiston Tribune*'s reporters, with the intention of using the photograph as part of a Christmas story in the paper. That man in the second photograph at the top of the page had a similar build, similar glasses, similar goatee, and similar clothing as the alleged thief shown in the video still at the bottom of the page.

When the *Lewiston Tribune* hit the streets on that Thursday, readers of the newspaper spotted the similarities between the two photographs almost immediately. Numerous phone calls by alert readers were made to the authorities pointing out the similarities. The police acted quickly and arrested the suspected thief. The two photographs were indeed of the same man, who was subsequently brought to justice.

Jami Johnson's wallet contained her driving license, three credit cards, and six hundred dollars. When the wallet was returned to her, the cash was unfortunately missing.[3]

4. On September 11, 2002, one year after the infamous terrorist attacks on the United States, the winning numbers in one of the New York State lottery drawings were 9, 1, 1. Over five thousand people had selected the winning numbers, each collecting 500 dollars.[4]
5. On the tenth anniversary of the 9/11 attacks, the first three winners at the Belmont Park Racetrack in New York had the numbers 9, 1, 1. Racing fans paid little notice on September 11, 2011, when the first race was won by the number 9 horse, *Say Toba Sandy*. But fans at the racetrack began to get excited when the second race was won by horse number 1, *Wishful Tomcat*. Could it be possible, they wondered, that the winner of the third race would also be a horse numbered 1, thereby completing the sequence 9, 1, 1? They did not have long to wait to find out. To the amazement of many fans at Belmont Park, horse number 1, *Haya's Boy*, came from behind and stormed home to victory in the third race. The 9, 1, 1, sequence had occurred, apparently, against incredible odds![5]
6. A surprising event occurred in a lottery in Bulgaria in 2009. On both September 6 and 10 of that year, a machine on live television selected from the numbers 1 to 42 the six numbers 4, 15, 23, 24, 35, 42. (In the second lottery draw, the numbers were in a different order.) The probability of this occurring is one in over 5.2 million. It was the first time that similar numbers had been drawn in the fifty-two-year-old history of the lottery in Bulgaria. The results caused so much suspicion among lottery players that Svilen Neikov, the Bulgarian sports minister, ordered an investigation into the matter. The investigation concluded that the drawing of similar lottery numbers in two consecutive draws was a freak coincidence![6]
7. There was an interesting news item on Irish television (*RTE*) on

Thursday, January 26, 2012. The news story concerned a totally white blackbird that was visiting a Dublin man's garden in Ballinteer, County Dublin, on a regular basis over the previous few weeks. The Dublin man took a number of photographs of the all-white blackbird in his garden. The man concerned said that he was neither a photographer nor a bird-watcher. BirdWatch Ireland was called in to give their opinion on the all-white blackbird. Their spokesman, Dick Coombes, said that the very rare specimen of blackbird was either an albino or leucistic. Incredibly, the man in whose garden the all-white blackbird regularly visited in Ballinteer was named *Robin* Cannon. The camera he used to photograph the all-white Blackbird was a *Canon*.[7]

8. Michael Dick, a carpenter from east London in the United Kingdom had split up with his first wife. Soon after, his daughter, Lisa, went to live in Sudbury, Suffolk, in England, with her mother.

   Ten years later, in 2007, Mr. Dick, now fifty-eight years old, and his two daughters from a second relationship, searched the streets of Sudbury, Suffolk, looking for his long-lost daughter. He looked up the local electoral register but still could not locate his daughter.

   He finally went to the Suffolk *Free Press* newspaper for help. They agreed to run the story and decided to photograph Mr. Dick, with his two daughters, Samantha aged twenty-two, and Shannon, aged ten, beside him on a public street in Sudbury. The newspaper published the photograph in its next edition.

   Mr. Dick's first daughter, Lisa, now aged thirty-one, discovered that her father was trying to locate her when friends mentioned that they had read the story in the Suffolk *Free Press* newspaper. Lisa read the story, looked at the photograph of her Dad and her two half-sisters, and almost immediately made contact with her father.

   Later Lisa looked at the newspaper photograph a little more carefully. Amazingly, she realized that both she and her mother were also in the photograph! They both appeared in the image about twenty yards behind her father and her two half-sisters.

   Lisa, who has three children of her own, lived in Colchester, in Essex, at the time. She explained to her dad that she had only been in Sudbury that day to visit her mom. Lisa and her dad were stunned by the coincidence that Lisa and her mother were walking by—within yards of her dad and two half-sisters—just as the photograph was taken.[8]

9. On February 9, 2012, there was an interesting case in the High Court in Dublin. It concerned a lengthy dispute between the Enniskerry Walkers Association (EWA) and a local landowner in the Glencree Valley, County Wicklow, in eastern Ireland (not far from Dublin) over an alleged right-of-way through the landowner's land. The Court ruled in favor of the landowner. Curiously, his name was Joseph Walker.[9]

Most people are surprised, amused, or even astonished when they come across coincidences of this nature. But should they be? The fact is we should expect strange coincidences to occur from time to time. Why? The explanation reaches to the very heart of the coincidence phenomena: On any given day of the week around the world, a *specified* rare event or coincidence is unlikely to occur; however, on any given day, an amazing, remarkable *unspecified* event is bound to happen somewhere.

The outstanding English mathematician John Edensor Littlewood (1885–1977) devised a neat method of illustrating that coincidences occur much more frequently than the average person realizes.

Littlewood defined a *truly surprising event* as one that has a probability of one in a million of happening. He assumed that the average person is awake for about eight hours per day. He also assumed that the average person heard or read or saw an event each second of her waking moments. Thus the average person will experience about thirty thousand events each day, or about a million per month. Consequently, the average person will experience a one-in-a-million event about once every month! This surprising result is now known by mathematicians around the world as *Littlewood's law*.[10]

The same point can be made by considering the population of a relatively large country. Take the United States, for example. There are over 300 million people living there. Consequently, a one-in-a-million coincidence should occur about 300 times every day in the United States. That is equivalent to more than 100,000 one-in-a-million chance coincidences every year in that country alone! It is easy to see from this that the number of such coincidences occurring around the world each day must be staggering. Not all of these remarkable coincidences are recorded. But those that are most often bewilder those who experience them, and indeed fascinate many of those who read of them.

Consider the following story, for instance. In 1997, a girl was born in St. Mary's Hospital, Portsmouth, England. The newborn girl was named Emily. (Her surname was Beard.) Emily was born on the 12th day of the 12th month at 12 minutes past noon. Her father, David, was born on the 4th day of the 4th month at 4:40 pm. Her mother, Helen, came into the world on the 10th day of the 10th month. Emily's brother, Harry, arrived in the world on the 6th day of the 6th month. Emily's paternal grandmother, Sylvia, made her first appearance in the world on the 11th day of the 11th month. The family only realized the string of coincidences after Emily's birth.[11]

What would any family make of these unusual coincidences?

This viewpoint that surprising events happen frequently is borne out by examining any large sample of random data. Although the data is arrived at random, patterns can be found within the data. A classic example of this phenomenon is the decimal expansion of pi. Because pi is an irrational number, its decimal digits go on forever without apparently conforming to any pattern.

Although it has not yet been proved, most mathematicians believe that pi is a *normal* number. A normal number is an irrational number in which each of the ten decimal digits occurs in the decimal expansion of the number the expected limiting number of times. For example, each of the ten digits would be expected to occur about one-tenth of the time; each of the pairs of digits 00 to 99 would be expected to occur about one-hundredth of the time. Each of the triplets of digits 000 to 999 would be expected to occur about one-thousandth of the time. And so on.[12]

However, within the decimal expansion of pi, patterns involving a *finite* number of digits are found. For example, the sequence 31415926 (which are the first eight digits of pi) can be found within pi beginning at decimal position 50,366,472.[13] If we go farther out the decimal sequence of pi, we find even more startling patterns. For instance, the ascending sequence 0123456789 can be first found beginning at decimal position 17,387,594,880. The descending sequence 9876543210 first occurs beginning at decimal position 21,981,157,633. So even in the midst of an apparent patternless and endless series of digits, patterns involving a *finite* number of digits can be found.[14]

This is to be expected, if the endless decimal expansion of pi is truly without pattern. Because the decimal expansion of pi is infinite, we could reasonably expect that all finite blocks of digits—in any order—will eventually be found in the never-ending series of decimals. For all we know, there may be some decimal position within pi such that the following billion digits are all nines! Or sevens! Or whatever digit you care to mention!

In fact, one would expect all finite blocks of digits—in any order—to exist an infinite number of times in the decimal expansion of pi, or indeed the decimal expansion of any irrational number! Thus any finite series of digits (however meaningful to human observers it may or may not be) must lie in wait, to be discovered, not once, but probably an infinite number of times, within the never-ending decimal expansion of all irrational numbers, including pi.

What is true of pi is true of the physical world, also. In a world where billions of different events happen daily, a small proportion of those events are going to appear surprising, even startling, to human observers! If they weren't, that would be surprising!

This way of looking at the world makes us realize that coincidences will and do occur more frequently than we might initially believe.

Does the fact that coincidences occur at all have any serious consequences? It appears that it does. There are many examples we could give that show how coincidence played a huge role in the existence of the human race. I would not be typing these words, nor would you be reading them, if it were not for the fact that my parents (and your parents also) met up and, for various reasons, decided to reproduce. Of course, my parents or yours would not exist if it were not for the fact that their parents got together and reproduced.

We can continue to reason like this and go back as far as we like, even to the time when our ancestors were fish swimming in the oceans, and even before that to the time when our ancestors were microbes in the early seas billions of years ago. If there was one slight change in the sequence of events from then to now concerning who reproduced and with whom, you and I would not be here, nor indeed would any of the other human beings that exist today.

But coincidence plays even a bigger part in the history of the universe. Cosmologists today believe that there was a coincidental collision between the very early Earth and another Mars-sized object that scientists have named Theia.[15] The debris, they believe, from the collision coalesced to form the moon. This theory is known as the *giant impact hypothesis*. Over the billions of years since that collision, the moon and its gravitational attraction has had an extremely stabilizing effect on the axis of Earth and stops it from wobbling like a spinning top. In other words, the moon has given Earth the stable environment for life (once life had begun) to flourish.[16]

There are scientists who will argue that the apparent benevolent role of coincidence goes beyond the moon's stabilizing effect on Earth. The giant planet Jupiter, they argue, is so big compared to Earth and other planets in the solar system that its gravitational pull attracts much debris—meteorites and the like—from space. This debris regularly crashes into the surface of Jupiter. If it were not for Jupiter, much of this debris would strike Earth, with devastating consequences for all forms of life here.[17]

Perhaps.

There is no doubt that Jupiter's gravitational pull attracts a lot of space junk. Some of that space debris may have hit Earth were it not for Jupiter, and some may have not. Some scientists argue that Jupiter does as much damage by deflecting asteroids in toward Earth as it does by throwing debris out of the solar system. So the question of whether Jupiter is blindly assisting life on Earth is far from being resolved and is, of course, very controversial. I am afraid that the jury is still out on this thorny question.[18]

Most scientists, however, have no doubt that it was a massive coincidental collision between Earth and a meteorite some sixty-five million years ago that wiped out 99 percent of all the species on our planet. This mass extinction cleared Earth of most of the large animals, including the dinosaurs, and provided an opportunity for a small species of mammal, about the size of a mouse, to flourish and reproduce. That species of mammal is a direct ancestor of the human race. If it were not for that meteorite strike in the far distant past, not only would you and I not be here, but the entire human race would most likely have never evolved![19]

This viewpoint that the human race exists because of a very unlikely coincidence is shared by many of the world's philosophers and scientists today. It is not a view that just became fashionable in recent years; the viewpoint has a long history.

Over the centuries, many have held the philosophical view that perhaps the universe itself came into existence because of a freak coincidence. Perhaps the quantum void that many scientists believe predated the universe, has always existed, but for one reason or another, unknown to us at the moment, its quantum laws have never produced a universe other than ours. Perhaps an extraordinary freak coincidence of how those laws behave in the quantum void caused the universe to burst into existence, creating equal amounts of positive and negative energy—thereby not violating any conservation laws of energy in the process. Perhaps that freak coincidence was such an unlikely possibility that it only ever happened once!

Or perhaps the quantum void is such that its quantum laws continuously produce an infinite number of universes. This may initially appear unlikely, but when you think about the matter, you are likely to wonder that if the quantum laws operating in the quantum void can produce one universe at one point in time, what is stopping those same laws from producing other universes or indeed, producing perhaps an infinite number of others? It appears (to me at least) that nature's laws are unlikely to be such that one universe *only* can emerge from the quantum void.[20]

Of course, as can be expected, coincidences are not always benevolent to the human race. One striking example of this has emerged over the last few years.

Two members of the physics department of the Texas State University–San Marcos issued interesting findings in 2012 concerning the sinking of the ill-fated RMS *Titanic*. (My hometown, Cobh, formerly known as Queenstown, was the last port of call for the *Titanic* before she sailed away into history on that April day in 1912.) Donald Olson and Russell Doescher from the Texas State physics department, along with Roger Sinnott, senior contributing editor at the astronomy magazine *Sky & Telescope*, published their findings in the April 2012 edition of the popular magazine.

It is well-known that the *Titanic* hit an iceberg on the night of Sunday, April 14, 1912. It is also known that despite wireless signals warning of icebergs in the shipping lanes ahead, the *Titanic* continued to steam relentlessly through the north Atlantic. The question that has baffled scientists for decades is this: Why were there icebergs that far south of Greenland that they would end up in the shipping lanes of transatlantic liners?

What Olson, Doescher, and Sinnott found was that there was an exceptionally close approach of the moon to Earth on January 4, 1912. The distance between Earth and the moon was just 221,440 miles. In fact, the moon's perigee—its closest approach to Earth—happened within six minutes of the full moon on January 4 of that year, and it was the closest the moon came to Earth in 1,400 years. In addition to all that, Earth's perihelion—its closest approach to the sun—occurred on January 3, 1912.[21]

The late oceanographer Fergus J. Wood, research associate, Office of the Director, National Ocean Survey, first published details in 1986 about that hypothesis concerning the abnormally high tide caused by the sun, moon, and Earth configuration in 1912.[22]

Getting back to 1912, the Earth–moon–sun configuration would have caused extremely high tides. Normally, in each spring icebergs tend to break apart off Greenland and head south. But such icebergs would not have been traveling fast enough to reach the north Atlantic shipping lanes by April.

Indeed, some icebergs become stranded and grounded in the shallow waters off the coast of Labrador and Newfoundland as they travel southward.

Olson, Doescher, and Sinnott argued in their paper that it was the unlikely celestial coincidence of the moon, Earth, and the sun that occurred on January 3 and 4, 1912, that caused many of these stranded and grounded icebergs off Labrador and Newfoundland to be freed up by the extremely high tides. Having been freed by the exceptionally high tides, these icebergs then traveled farther southward and would have reached the shipping lanes of the north Atlantic by April. It was *Titanic*'s fate to collide with one such iceberg 375 miles south of Newfoundland; and the rest, as they say, is history.

Soon after reading the above article in *Sky & Telescope*, I played around with a few numbers and discovered the following little curiosity:

In 1912, the moon is closest to Earth for 1,400 years	1400
In that year, the *Titanic* sinks 375 miles south of Newfoundland	375
The *Titanic* hits an iceberg in the 136th month of the century	136
The *Titanic* is the number 1 ocean cruise liner in the world	+ 1
Total equals the year that the *Titanic* sank	1912

Figure 15.1. Strange Number Curiosity concerning the RMS *Titanic*

Incidentally, here is a little table concerning the *Titanic* that readers will not have come across before:

Length (in whole feet) of the *Titanic*	882
Width (in whole feet)	92
Number of persons who boarded the *Titanic* at Queenstown (Cobh)	123
Shipyard number of the *Titanic*	401
Date the *Titanic* struck the iceberg (4/14)	+ 414
Total equals year that the *Titanic* sank	1912

Figure 15.2. Another Strange Number Curiosity concerning the RMS *Titanic*

Here is an interesting way you can illustrate to friends how surprising events occur much more frequently than most people are inclined to believe. Suppose you are at a gathering consisting of twenty-three people. Suppose someone asks for each of the birthdays (date and month, ignoring the year) of all twenty-three people. To everyone's surprise, it is found that two birthdays match. Is it a remarkable coincidence?

If you are familiar with the laws of probability, you will not be too surprised. Why? Because the laws of probability tell us that in a group of twenty-three people, the probability that at least two of them share a birthday (date and month) is 50 percent.[23] (This calculation ignores birthdays falling on February 29.) If there are thirty people at the gathering, the probability that at least two of the thirty persons will share a birthday rises to a little over 70 percent.

With access to the Internet, it may prove worthwhile to go online at www.WolframAlpha.com. The information that can be found there will give solutions to this type and many other types of problems.[24]

There are different mathematical methods of obtaining the answer to problems such as these. Using a scientific calculator, we can solve a problem such as this in the following manner: Let's assume that there are thirty people at a gathering. What is the probability that two of them will share a birthday? (Once again, this calculation ignores birthdays falling on February 29.) First, perform the following calculation to obtain the probability that there will NOT be a match:

$$\frac{365P30}{365^{30}} = 0.29368375\ldots.$$

(The symbol P represents permutations. Thus $365P30$ means the number of permutations of 365 objects, taken 30 at a time.) Divide $365P30$ by 365^{30} on your calculator. Subtract the answer from 1 (certainty) to find the probability that there will be a match: $1 - 0.29368375$. This equals 70.631624 percent that there will be a match.

A bet based on this fact can be a handy little earner. If you are at a party consisting of thirty people or more, get every one of them to write down their birthday (date and month only). Bet that at least two people's birthdays will match. If there are thirty people at the party, you can expect to win this bet about seven times out of every ten. If there are thirty-five people at the gathering, the probability increases to about 81.43 percent that there will be a match; if 39 people are at the party, the probability increases to about 87.82 percent.

The same laws of probability apply to similar situations. Suppose you ask a group of nine people to write the name of any state of the United States. What is the probability that two people will write the name of the same state?

Surprisingly, it is 53.45 percent. So it is better than evens that there will be a match.

The same method of obtaining the answer as given above can be used to solve this problem. The formula to obtain the probability that there will NOT be a match is:

$$\frac{50p9}{50^9} = 0.46549595 \ldots.$$

Therefore, the probability that there will be a match is 1 – 0.46549595 . . . . This equals 53.450404 . . . percent that there will be a match.

If the same question is put to a group consisting of fifteen people, the probability is found to be over 90.35 percent that there will be a match. With twenty-two people, the probability rises to 99.58 percent that there will be a match. (This essentially is the same problem as asking what the probability is if nine people, or fifteen people, or twenty-two people, chosen at random, are asked to pick a date [month and day] from fifty possible dates.)

These examples illustrate once again that some events that appear to be highly unlikely are actually more likely than not to occur.

This surprising fact can help you to pick up some loose change from an unsuspecting mark. Propose the following bet to him. Tell him that you are about to throw a pair of six-sided fair dice, but before you do so you are going to give the mark a chance to earn a few dollars. Ask him to name two numbers from 1 to 6 inclusive. Suppose the mark names 1 and 3. You now bet the mark that the numbers 1 or 3 will show on at least one of the dice that you throw. Which of you is favored by the odds?

The mark will probably reason as follows: This guy (you!) is about to throw two six-sided dice. He has asked me to name two numbers from 1 to 6 inclusive. I named 1 and 3. He bets me that one of those numbers will turn up on the roll of the two dice. But there are two dice, each with six numbers. Therefore the odds must favor me.

Of course that is exactly how you want the mark to reason. His reasoning, however, is incorrect!

Surprisingly, the odds favor you. The chances of the numbers 1 or 3 (or any two numbers from 1 to 6) NOT coming up on one roll of one die is 4/6. Therefore, the chances of 1 or 3 NOT coming up on one roll of two dice is 4/6 · 4/6, which equals 16/36. Thus there are 16 chances out of 36 that 1 or 3 will

NOT turn up. Therefore, there are 20 chances in 36 that 1 or 3 will turn up! This reduces the probability in your favor to 5 chances in 9. Or to say the same thing another way, the odds are 5 to 4 in your favor!

When a coincidence occurs, it is not always possible to use the theory of probability to establish the likelihood of the coincidence. Why? Because there may be simply too many unknown variables. For example, what is the likelihood of rain falling in Rio de Janeiro, in London, and in Tokyo on four consecutive Mondays in January in the same year? There are just too many unknowns to give any accurate answer. When probability theory cannot be used to give the exact probability of a coincidence occurring, it just heightens the mystery in the minds of many who wonder why the coincidence occurred at all.

This is why, down through the years, the public has been fascinated with coincidences. Newspaper and magazine editors have, of course, long known this, and from time to time they publish a remarkable coincidence that grabs the public's attention. Consider, for example, the following case. On August 28, 1974, Great Britain's ex–prime minister Edward Heath helped publicize a novel written by John Dyson titled *The Prime Minister's Boat Is Missing*.[25] Eight days later, on September 5, 1974, Edward Heath's yacht, *Morning Cloud* 111, was lost at sea.[26]

Another strange example of coincidence showing its face concerns the case of the thirty-one-year-old twin sisters Lorraine and Levinia. They both lived in different villages, in East Anglia, England. Both twins were driving to each other's homes on Christmas Eve in 1994. Both of the twins cars collided head-on into each other on a single-track country lane in icy conditions, at Flitcham, Norfolk, in England. They were both brought to Elizabeth Hospital, in King's Lynn, suffering from whiplash injuries and concussion.

By a strange coincidence, those twin sisters were named Lorraine and Levinia Christmas.[27]

Getting back to coincidence and its appearances on this little planet as it orbits a star about ninety-three million miles away, we have learned that we should expect coincidences to happen every now and then. But when an exceptionally rare event does occur, our minds find it extremely difficult to come to terms with an event that we view as truly surprising.

Consequently, when an egg-sized meteorite smashed through the roof of a family home in a suburb of Paris, in France, in the summer of 2011, people became excited. The family in question was away on vacation at the time. When they returned, they noticed that their roof was leaking and hired someone to fix it. That individual was amazed at the way a tile on the roof of the house was apparently violently demolished, and he speculated that only an object falling from the sky could have caused such damage. Eventually, the family found the culprit that did the damage: an egg-sized, dark-colored rock weighing about three and a half ounces.

The family called in a scientist and mineral expert named Alain Carrion to investigate the piece of rock. The homeowners were amazed to hear that this rock had come from the belt of asteroids that lay between Mars and Jupiter, and that it was over four and a half billion years old! The rock probably contained dust grains that existed before the solar system, including Earth, was formed! This means that these dust grains came from outside our solar system.

Alain Carion told news reporters that over the last four centuries, it is believed that only about fifty meteorite strikes had hit France.

The family in question must be still wondering why, of all the places on Earth this meteorite should fall, it fell on their house. The fact that the meteorite should fall on dry land at all is unusual in itself, because only about 30 percent of Earth's surface is dry land. So for it to fall on someone's house is certainly an unlikely event.

What makes this story even more unlikely is that the name of the family whose roof the meteorite smashed through is *Comette*![28]

But meteorites falling toward Earth have to fall somewhere. Tens of thousands of them have fallen over the centuries. But every now and again, something unusual will occur when a meteorite falls. That is what happened to the Comette family. Of course, if the meteorite hit some other house, be it in Paris or in some other location on Earth, the occupants would most probably be asking, why was it their house that was struck?

I sent this chapter to Dr. Cong, asking him if he had any unusual coincidences to share with us. He is still living on the West Coast of the United States. I am conscious of the fact that he pretends to believe in the occult power of numbers. At least I think he pretends! I think he projects this image so as to impress many of his admirers.

One can never be certain of what this jokester believes. However, Dr. Cong is no fool. He is well acquainted with the properties of numbers, and I thought it would be interesting to see what he had to say about coincidences.

He sent back the following e-mail.

Hi, Owen,

Thank you for sending your chapter on coincidences. As soon as I saw it, I realized how lucky we are to have the benefit of e-mail.

Like you, I have a number of these strange occurrences in my files. But I have a few others that are completely original with me. You will not read any of these little discoveries of mine in any book on coincidences. I will give you a few of them for your book and hope that your readers will enjoy them.

The first concerns the attack by Japan on Pearl Harbor on December 7, 1941, which brought the United States into the Second World War. Consider the following statement:

JAPANESE ATTACK USA.

Now use the following alphabet code where A = 1, B = 2, C = 3, and so on. Substitute those letters in the three words above. Then add up the values of the letters in the two words JAPANESE ATTACK and place them underneath those two words. Then add up the values of the letters USA and place that value underneath those three letters. This is what you will get:

JAPANESE ATTACK	USA
127	41

Curiously, the Japanese attack on the United States at Pearl Harbor occurred on 12/7/41.

Here's another, similar, coincidence. One of the most beloved US presidents is, of course, Abraham Lincoln. He was born on Sunday, February 12, 1809. Lincoln was the sixteenth president of the United States. He was shot by John Wilkes Booth on Friday, April 14, 1865. He died in the early hours of the following morning. I spotted this coincidence concerning this most beloved president when I was just twelve years old.

Use the same alphabet code as above, and substitute those numbers in the following words:

SIXTEENTH PRESIDENT   A. LINCOLN

Now obtain the sum of the letters in the two words SIXTEENTH PRESIDENT. The sum equals 234. The sum of the letters in the name A. LINCOLN equals 80. If we concatenate the words above, we obtain the phrase SIXTEENTH PRESIDENT. A. LINCOLN. If the two numbers 234 and 80 are concatenated, we obtain the number 23,480. By a remarkable coincidence, President Lincoln was shot on the 23,480th day of the nineteenth century.

The following number curiosity involving President John F. Kennedy is worth noting as well. JFK, as he was widely known, was also one of the most beloved presidents of the United States. He is remembered to this day by millions of people around the world.

John F. Kennedy was born on Tuesday, May 29, 1917, and was assassinated on Friday, November 22, 1963. The world was stunned and saddened when he was struck down on that fateful day. It has been said that most people who were alive on that November day long ago remember where they were and what they were doing when the tragic news came through that the president of the United States of America had been assassinated.

Now to the curiosity. Consider the following two phrases: DEATH OF PRESIDENT and KENNEDY. Using the usual alphabet code above, the sum of the letters in the phrase DEATH OF PRESIDENT equals 169 and the sum of the letters in the word KENNEDY equals 78. When the four words are concatenated, we obtain the phrase DEATH OF PRESIDENT KENNEDY. When the two numbers 169 and 78 are concatenated, we get the number 16,978. John F. Kennedy was exactly 16,978 days old when he was assassinated.

Here's a little set of coincidences concerning JFK and the number 35. This number (or its digits) apparently entered John F. Kennedy's life in a number of ways. We know his date of birth was Tuesday, May 29, 1917. He was thus born on the 3rd day of the week and on the 5th Tuesday of the month. He was born on the 149th day of the year. It so happens that 149 is the 35th prime number. Consider the two words THIRTY-FIFTH. Using the earlier alphabet code, the sum of the letters in the words THIRTY-FIFTH equals 149—again the 35th prime shows up.

John F. Kennedy had 3 brothers and 5 sisters. He was assassinated in the 35th month of the decade. The initials of his reputed assassin, Lee Harvey Oswald, are the 12th , 8th, and 15th letters of the alphabet. Those three numbers sum to 35. The date March 5, 1953, may be written as 3/5/53. Note the digits of this palindrome date! On the 3/5/53, John F. Kennedy was exactly 35 years, 35 weeks, and 35 days old.

JFK was married on September 12, 1953. The last two digits are 35 in reverse. He was elected president on November 8, 1960, just 53 days before the end of the year. John F. Kennedy was assassinated on 11/22/1963. Note that $11 - 2 - 2 + 19 + 6 + 3 = 35$. One year after his assassination, his widow,

Jackie, went to live on the 15th floor of 1040 Fifth Avenue, in New York City. Note that 15 equals $3 \cdot 5$ and $1040 = 1122 - 19 - 63$. This is the date that John F. Kennedy was assassinated. Finally, John F. Kennedy was the 35th president of the United States.

The two following curiosities concerning Neil Alden Armstrong, the first man to walk on the moon, are also interesting. Armstrong was born in Ohio on August 5, 1930, and he died on August 25, 2012. He landed on the moon on July 20, 1969. That was the 201st day of the year. Using our alphabet code, the sum of the letters in the name NEIL ALDEN ARMSTRONG is 201.

Neil Armstrong was one of three astronauts involved in one of the most famous and memorable events in human history when he walked on the moon on July 20, 1969. Millions around the globe watched grainy pictures on television as Armstrong stepped on to the lunar surface to become the first man to walk on the moon.

As I said above, Armstrong was born on August 5, 1930, and he died on August 25, 2012. Using the phrase NASA ASTRONAUT ON MOON it translates into 250, using our alphabet code. Armstrong was 38 years old when he landed on the moon. When we concatenate the two numbers 250 and 38, we get 25,038. Neil Armstrong landed on the moon on Sunday, July 20, 1969 (Houston Time). That day was the 25,038th day of the twentieth century!

Here are some number coincidences concerning a famous moment in history. The atomic bomb has been used, to date, against only one country, Japan. Using our code, the sum of the letters in the word JAPAN equals 42.

The great Italian physicist Enrico Fermi and his wife, Laura, who was Jewish, came to the United States in 1938, to escape anti-Jewish laws in Italy. He was aged 41 when he and his science team achieved the first nuclear chain reaction on December 2, 1942, in the squash court underneath the West Stand of Stagg Field, at the University of Chicago. It is curious that Fermi made the remarkable breakthrough in nuclear physics when he was in his 42nd year, with 42 scientists beside him, in the year '42, on the $4 \cdot 2 \cdot 42$ day of the year.[29]

Enrico Fermi succeeded in achieving fission with Uranium 235 on December 2nd. The sum of the letters in the words DECEMBER SECOND and ENRICO FERMI both equal 115. Note that 23 times 5 equals 115.

Fermi was born on 9/29/01. He was aged 41 when—on the 2nd day of the last month of the year 1942—he achieved the self-sustaining atomic reaction. It is curious that $(929 + 01 + 41) \cdot 2 = 1942$. The date of the scientific breakthrough may be written as 12/2/1942. It is curious that $12^2 + 194 - 2 = 336$, which equals the day of the year the breakthrough was made.

Coincidences do not just occur in the lives of famous people. They happen to people in all walks in life.

For example, the not-so-long arm of coincidence made an appearance in the case of American Airlines flight number 587, which took off from Kennedy Airport in New York at 9:14 a.m. local time (14.14 Coordinated Universal Time) on Monday, November 12, 2001. It crashed into a quiet Queens's neighborhood soon after takeoff. All 260 people aboard the Airbus A300 were killed in the crash. (Five people on the ground were also killed.) The air disaster—occurring just less than nine weeks after the 9/11 attacks—rocked New Yorkers, who had had initial fears that the plane had been a target for terrorists.

That afternoon, thousands of people purchasing tickets in the New Jersey Lottery Pick-3 game bet on the numbers 5, 8, and 7 (the flight number). Those numbers were duly drawn that evening as the winning numbers, resulting in over $1 million being paid out!

The Pick-3 is a daily lottery game in which players are invited to pick any three-digit number from 000 to 999. There are one thousand different three-digit numbers in that range. Therefore, the probability of picking one three-digit number in the Pick-3 game is one in a thousand.

To make the story even more remarkable, it should be noted that there are two daily draws in the New Jersey Pick-3 lottery, one shortly after midday, and the other later in the evening. November 12, 2001, was the first day that two daily draws were held. Amazingly, the numbers drawn in the lottery shortly after midday on that very same day, November 12, 2001, were 5, 7, and 8. This is the number 587 in a different cyclic order!

I devised the table below on the following day, which I have never before published:

A Curiosity concerning Flight 587
by Dr. Cong

Name of Flight 587	587
Coordinated Universal Time of takeoff from JFK	+ 1414
Total equals the year that Flight 587 crashed	2001

Figure 15.3

Readers may have read of many coincidences that refer to the dreadful events of 9/11, in 2001. But here is one curiosity readers will not have heard of before.

Consider the following sentence.

Nine eleven two thousand and one was the date that the Twin Towers in America were destroyed.

Using our alphabet code from above, the sum of the letters in that sentence equals 911.

Here is a curiosity concerning the statistics of One World Trade Center, which was built on the site of the former Twin Towers in New York City.

Height of One World Trade Center (in whole feet)	1776
Height of World Trade Center (in whole meters)	541
Area of site of the World Trade Center (in acres)	16
Number of States in the United States	+ 50
Total	2383

The 2,383th prime number is 21,191.

Curiously,

21	19	1
U	S	A

The 21st letter of the alphabet is U; the 19th letter is S; and the 1st letter is A.

I mentioned that the height of One World Trade Center in meters is 541. It is curious that 541 is the $(9 \cdot 11 + 01)$th prime number. Those digits give the date of the (9/11/01) attack.

When major atrocities or accidents occur, people often look for any patterns that may be connected to the event. I suppose it is an attempt by many people to make sense of the terrible suffering that is inflicted on so many innocents as a result of these tragic events. Consider the following terrible event that unfolded on the East Coast of the United States in the late 1950s.

On Monday, September 15, 1958, a New Jersey commuter train, numbered 3314, crashed through a drawbridge and plunged into Newark Bay. There were 48 people killed in the tragedy. That night, television pictures showed the last passenger car being hoisted from the Bay. Prominently displayed in the middle of the picture was the car number, 932. The following day, thousands of punters playing the Manhattan numbers game bet on the number 932 and won! It has been claimed that consequently $50 million was paid out in the New York– New Jersey area.

It is curious that the car number turned out to be 932. You see, the original bridge across Newark Bay was built in 1864. One-half of that number is 932. Many years ago, when I first read of this disaster, I put together a little table concerning these tragic events. Have a look at the first table I devised:

The last passenger rail car hoisted from the Bay was car number 932.

$932 + 932 = 1864$ = Year that the original Newark Bay Bridge was built.

$932 + 932 + 9^2 + 3^2 + 2^2 = 1958$ = Year of Newark Bay Railroad Disaster.

Figure 15.4. First Curiosity relating to Newark Bay—New Jersey Rail Disaster, 1958.

The fact that the train was numbered 3314 is also curious, given the following set of figures.

Year of Newark Bay Railroad Disaster	1958
Month number in which railroad accident occurred	9
Date of month in which railroad accident occurred	15
Day number of year in which railroad accident occurred	258
Number of people killed in the railroad accident	48
Number of years after original bridge was built in which accident occurred	94
Number of last railroad car taken from water	+ 932
Total = Number of Railroad Train involved in accident	3314

Figure 15.5. Second Curiosity relating to Newark Bay/New Jersey Rail Disaster, 1958.

The appearance of the numbers 3314 and 932 in the Newark Bay Rail Disaster is also curious for the following reason: The number of prime numbers less than 3,314 is 466. (The 466th prime number is 3,313.) Curiously, 466 + 466 equals 932.

Sometimes truth is stranger than fiction!

I remain,

Your faithful friend,

Dr. Cong

CHAPTER 16

BEAUTIFUL MATHEMATICAL EQUATIONS

How do we describe mathematical beauty? The question has been pondered by many mathematicians and philosophers for centuries. I think most mathematicians would say that a beautiful mathematical equation is something that states a great amount of information in as few concise terms as possible; something that brings home to the individual gazing on the equation that there is a wonderful order behind things; something in the result that shows that the particular result had to be—that the truth of that particular proposition could not be different.

In this chapter I hope to illustrate some of the beauty that is to be found in mathematics by showing the reader a number of simple patterns found in numbers. These patterns have been described by some of the greatest minds on Earth as being beautiful.

The great American physicist Murray Gell-Mann (1929–), who is a self-proclaimed agnostic,[1] often spoke of the mystery that beautiful equations in science are more likely to be true than ugly equations. He went on to say that a beautiful proof can be defined as one that expresses a theory in a very short space, without unnecessary complications.

Gell-Mann also asked the question: "Why is beauty or elegance a successful criterion in choosing a correct theory in fundamental physics?"[2] No one knows the answer to this question. The best we can say is that it appears that nature—at its very core—is beautiful, and thus the equations expressing the laws of nature are also beautiful.

The beauty in mathematical equations has been noticed for centuries by scientists, philosophers, and, of course, mathematicians. The outstanding Hungarian mathematician Paul Erdős (1913–1996) often spoke of an imaginary "Book," in which God had written all the most beautiful proofs in mathematics. When Erdős encountered a particularly attractive and elegant proof, he would express his appreciation by saying: "This proof is from the Book." Of course, Erdős was an agnostic, and he certainly did not literally believe in the existence of such a "Book."[3]

By referring to the "Book," the Hungarian genius was putting forward his *platonic* view relating to the reality of mathematics in a poetic manner. Saying the same thing another way, Erdős was putting forward his belief that the truths of mathematics are eternal and unchanging, and that when we come across a mathematical proof, we *discover* it rather than *invent* it.

Let us look at some beautiful patterns that we observe when we look at these mysterious and amazing entities. The patterns that I give here are a small sample from a collection I have gathered over the years, and they certainly are not exhaustive.

Let's begin our gaze on some of these beautiful patterns by considering two simple patterns involving the squares of integers, commencing with the square of 1. Consider the series of odd numbers: 1, 3, 5, 7, 9, 11, 13, 15, 17, 19, 21, 23 . . . . Note the following pattern:

$$1 = 1^2$$
$$1 + 3 = 2^2$$
$$1 + 3 + 5 = 3^2$$
$$1 + 3 + 5 + 7 = 4^2$$
$$1 + 3 + 5 + 7 + 9 = 5^2$$
$$1 + 3 + 5 + 7 + 9 + 11 = 6^2$$
$$\ldots$$

Figure 16.1

If we denote the number of terms on the left-hand side of any row of the equation sign in figure 16.1 as n, the number on the right-hand side equals n^2.

The second pattern is very simple, but is also surprising:

$$1 = 1^2$$
$$1 + 2 + 1 = 2^2$$
$$1 + 2 + 3 + 2 + 1 = 3^2$$
$$1 + 2 + 3 + 4 + 3 + 2 + 1 = 4^2$$
$$1 + 2 + 3 + 4 + 5 + 4 + 3 + 2 + 1 = 5^2$$
$$\ldots$$

Figure 16.2

We encountered triangular numbers earlier in this book. You may recall that the triangular numbers are formed by the following procedure:

$$1 = 1$$
$$1 + 2 = 3$$
$$1 + 2 + 3 = 6$$
$$1 + 2 + 3 + 4 = 10$$
$$1 + 2 + 3 + 4 + 5 = 15$$
$$\ldots$$

Figure 16.3

The formula for triangular numbers is $n\,\frac{n+1}{2}$, where n is a positive integer. The series of triangular numbers, commencing with 1, is 1, 3, 6, 10, 15, 21, 28, 36, 45, 55, 66, 78 . . . .

The following pattern reveals a relation between the triangular numbers and the cubic numbers.

$$1 \cdot 0 + 1 = 1^3$$
$$3 \cdot 2 + 2 = 2^3$$
$$6 \cdot 4 + 3 = 3^3$$
$$10 \cdot 6 + 4 = 4^3$$
$$15 \cdot 8 + 5 = 5^3$$
$$21 \cdot 10 + 6 = 6^3$$
$$28 \cdot 12 + 7 = 7^3$$
$$\ldots$$

Figure 16.4

There are some beautiful patterns involving the triangular numbers in chapter 11. Let us look upon two other patterns involving the triangulars here. Consider the following patterns as shown figures 16.5 and 16.6:

$$1 \cdot 3 = (2 \cdot 1^2)^2 - 1^2$$
$$6 \cdot 10 = (2 \cdot 2^2)^2 - 2^2$$
$$15 \cdot 21 = (2 \cdot 3^2)^2 - 3^2$$
$$28 \cdot 36 = (2 \cdot 4^2)^2 - 4^2$$
$$45 \cdot 55 = (2 \cdot 5^2)^2 - 5^2$$
$$66 \cdot 78 = (2 \cdot 6^2)^2 - 6^2$$
$$\ldots$$

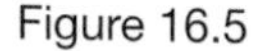

Figure 16.5

$$
\begin{aligned}
1^2 + 3^2 &= T_{(2\cdot 2)} \\
6^2 + 10^2 &= T_{(4\cdot 4)} \\
15^2 + 21^2 &= T_{(6\cdot 6)} \\
28^2 + 36^2 &= T_{(8\cdot 8)} \\
45^2 + 55^2 &= T_{(10\cdot 10)} \\
66^2 + 78^2 &= T_{(12\cdot 12)} \\
91^2 + 105^2 &= T_{(14\cdot 14)} \\
&\dots
\end{aligned}
$$

Figure 16.6

Undoubtedly the pattern shown in figure 16.7 must rank as one of the most beautiful patterns known that involves the triangular numbers. The identity captures the beautiful simplicity and elegance of the triangular numbers in a surprising manner.

$$
\begin{gathered}
T_1 + T_2 + T_3 = T_4 \\
T_5 + T_6 + T_7 + T_8 = T_9 + T_{10} \\
T_{11} + T_{12} + T_{13} + T_{14} + T_{15} = T_{16} + T_{17} + T_{18} \\
T_{19} + T_{20} + T_{21} + T_{22} + T_{23} + T_{24} = T_{25} + T_{26} + T_{27} + T_{28} \\
T_{29} + T_{30} + T_{31} + T_{32} + T_{33} + T_{34} + T_{35} = T_{36} + T_{37} + T_{38} + T_{39} + T_{40} \\
T_{41} + T_{42} + T_{43} + T_{44} + T_{45} + T_{46} + T_{47} + T_{48} = T_{49} + T_{50} + T_{51} + T_{52} + T_{53} + T_{54} \\
\dots
\end{gathered}
$$

Figure 16.7

The odd integers, and also the cubic numbers, appear in the next simple but beautiful pattern:

$$
\begin{gathered}
1 = 1^3 \\
3 + 5 = 2^3 \\
7 + 9 + 11 = 3^3 \\
13 + 15 + 17 + 19 = 4^3 \\
21 + 23 + 25 + 27 + 29 = 5^3 \\
31 + 33 + 35 + 37 + 39 + 41 = 6^3 \\
43 + 45 + 47 + 49 + 51 + 53 + 55 = 7^3 \\
\dots
\end{gathered}
$$

Figure 16.8

The following pattern illustrates a beautiful series involving the consecutive natural numbers, commencing with 1. The pattern, which is not well known among nonmathematicians, illustrates the simple and elegant beauty that is to be found at the very heart of mathematics.

$$1 + 2 = 3$$
$$4 + 5 + 6 = 7 + 8$$
$$9 + 10 + 11 + 12 = 13 + 14 + 15$$
$$16 + 17 + 18 + 19 + 20 = 21 + 22 + 23 + 24$$
$$25 + 26 + 27 + 28 + 29 + 30 = 31 + 32 + 33 + 34 + 35$$
$$36 + 37 + 38 + 39 + 40 + 41\ 42 = 43 + 44 + 45 + 46 + 47 + 48$$
$$\ldots$$

Figure 16.9

There is a similar beautiful pattern involving the integers raised to the power of 2. Those with an inquiring mind who look on the pattern in figure 16.10 must surely be moved by the way the sums of the square numbers in the pattern fit together like the pieces of a jigsaw puzzle:

$$3^2 + 4^2 = 5^2$$
$$10^2 + 11^2 + 12^2 = 13^2 + 14^2$$
$$21^2 + 22^2 + 23^2 + 24^2 = 25^2 + 26^2 + 27^2$$
$$36^2 + 37^2 + 38^2 + 39^2 + 40^2 = 41^2 + 42^2 + 43^2 + 44^2$$
$$55^2 + 56^2 + 57^2 + 58^2 + 59^2 + 60^2 = 61^2 + 62^2 + 63^2 + 64^2 + 65^2$$
$$78^2 + 79^2 + 80^2 + 81^2 + 82^2 + 83^2 + 84^2 = 85^2 + 86^2 + 87^2 + 88^2 + 89^2 + 90^2$$
$$\ldots$$

Figure 16.10

Note that the integers closest to the consecutive equation signs in figure 16.10 are two legs of Pythagorean triples. For example, we find in the first row, $3^2 + 4^2 = 5^2$. In the second row, we find 12^2 and 13^2. In the third row, we find 24^2 and 25^2. And so on.

Figure 16.11 shows a beautiful equation (involving the ten integers on the left-hand side of the equation sign). To obtain this equation I slightly modified a set of equations given by the late Martin Gardner in his book *The Magic Numbers of Dr. Matrix*.[4] In this version of the identity, each of the ten digits appear on the left-hand side of the equation. The right-hand side is simply 1.

$$\frac{12}{3 \cdot 4} \cdot \frac{56}{7 \cdot 8} \cdot 9^0 = 1$$

Figure 16.11

The set of equations shown in figure 16.12 has, since childhood, always struck me as being especially beautiful and elegant:

$$1^2 = 1^3$$
$$(1+2)^2 = 1^3 + 2^3$$
$$(1+2+3)^2 = 1^3 + 2^3 + 3^3$$
$$(1+2+3+4)^2 = 1^3 + 2^3 + 3^3 + 4^3$$
$$(1+2+3+4+5)^2 = 1^3 + 2^3 + 3^3 + 4^3 + 5^3$$
$$(1+2+3+4+5+6)^2 = 1^3 + 2^3 + 3^3 + 4^3 + 5^3 + 6^3$$
$$(1+2+3+4+5+6+7)^2 = 1^3 + 2^3 + 3^3 + 4^3 + 5^3 + 6^3 + 7^3$$
$$\ldots$$

Figure 16.12

The following pattern, involving division, is also unexpected and pretty:

$$(1+2)/3 = 1$$
$$(4+5+\ldots+10+11)/12 = 5$$
$$(13+\ldots+37+38)/39 = 17$$
$$(40+\ldots+119)/\ 120 = 53$$
$$(121+\ldots+362)/363 = 161$$
$$(364+\ldots+1091)/1092 = 485$$
$$\ldots$$

Figure 16.13

The numbers on the right-hand side of the equations are, 1, 5, 17, 53, 161, 485 . . . . These numbers in order satisfy the equation $2 \cdot 3^n - 1$, when n equals 0, 1, 2, 3, 4, 5 . . . .

Philippe Deléham, from Vouziers in northern France, contributed the following curiosity[5] about the series of numbers 1, 5, 17, 53, 161, 485, to the On-line Encyclopedia of Integer Sequences, on February 23, 2014:

$$1 = 1$$
$$1 + 3 + 1 = 5$$
$$1 + 3 + 9 + 3 + 1 = 17$$
$$1 + 3 + 9 + 27 + 9 + 3 + 1 = 53$$
$$\ldots$$

Figure 16.14

Here is one more simple but beautiful set of equations:

$$0 + 1 = 0^3 + 1^3$$
$$2 + 3 + 4 = 1^3 + 2^3$$
$$5 + 6 + \ldots + 8 + 9 = 2^3 + 3^3$$
$$10 + 11 \ldots + 15 + 16 = 3^3 + 4^3$$
$$17 + 18 + \ldots + 24 + 25 = 4^3 + 5^3$$
$$26 + 27 + \ldots + 35 + 36 = 5^3 + 6^3$$
$$\ldots$$

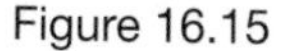

Figure 16.15

Besides simple number equations, there are of course many beautiful equations involving π. The following,[6] I think, is particularly simple and elegant:

$$\pi = 3 + \frac{4}{2 \cdot 3 \cdot 4} - \frac{4}{4 \cdot 5 \cdot 6} + \frac{4}{6 \cdot 7 \cdot 8} - \frac{4}{8 \cdot 9 \cdot 10} + \frac{4}{10 \cdot 11 \cdot 12} \ldots$$

Figure 16.16

There are also beautiful equations involving *e* and the square root of negative 1. One such equation reveals the astonishing fact that the so-called imaginary number *i*, which equals $\sqrt{-1}$, when raised to the power of itself, is equal to a positive number! It equals 0.207879576 . . .

$$i^i = \left(\frac{1}{\sqrt{e}}\right)^{\pi} = 0.207879576 \ldots$$

Figure 16.17

The fact that the two famous transcendental numbers, π and *e*, appear in the equation is also astonishing.

Many professional mathematicians believe the following equation is the most beautiful in the whole of mathematics. But it is just not professional mathematicians who are impressed by this equation, which was discovered, incidentally, by Leonhard Euler.

Euler's wonderful discovery has often been voted the most beautiful equation in mathematics in many surveys throughout the world. In a poll in 1990, the mathematics magazine *Mathematical Intelligencer* voted it the most beautiful in all of mathematics.[7]

The equation involves the mathematical operations addition, multiplication, and exponentiation. It also includes the five most famous identities in mathematics, 0, 1, e, π, and $\sqrt{-1}$. Here's the equation[8]:

$$e^{i\pi} + 1 = 0$$

Figure 16.18

The equation is truly astonishing. If you rewrite it as follows, you also discover another astonishing fact:

$$e^{i\pi} = -1$$

Figure 16.19

In our discussion of the number e in chapter 13, we discussed the natural logarithm of a number. For example, the natural logarithm of, say, 3, is 1.098612288 . . . because $e^{1.098612288\ldots} = 3$.

In other words, in the equation $e^x = y$, the exponent x is seen to be the natural logarithm of y. Thus in figure 16.19, the exponent $i\pi$ is the natural logarithm of -1. We write this result as shown in figure 16.20:

$$i\pi = \ln(-1)$$

Figure 16.20

Thus $i\pi$ is the natural logarithm of -1. We know that $-i$ equals i^3. Therefore, if we multiply both sides of the equation in figure 16.20 by $-i$, the answer must equal

$$i^4\pi = -i \cdot \ln(-1)$$

Figure 16.21

Because i^4 equals 1, the equation in figure 16.21 gives the following astonishing result[9]:

$$\pi = -i \cdot \ln(-1)$$

Figure 16.22

Who, in their sane mind, would have thought that the transcendental number 3.14159265 . . . would equal $-i$ times the natural logarithm of -1, as shown in figure 16.23?

$$3.14159265\ldots = -i \cdot \ln(-1)$$

Figure 16.23

Let us take another look at the remarkable equation in figure 16.19. Since i^2 equals -1, we insert i^2 in place of -1 in the equation. The equation then is

$$e^{i\pi} = i^2$$

Figure 16.24

We now raise both sides of the equation in figure 16.24 to the power of $-i$. Since $-i$ equals i^3, this gives

$$e^{i\,i\,i\,i\,\pi} = i^{-2i}$$

Figure 16.25

Since $i \cdot i \cdot i \cdot i = i^4 = 1$, we conclude that the equation in figure 16.25 equals the astonishing result[10] shown in figure 16.26:

$$e^{\pi} = i^{-2i}$$

Figure 16.26

Since e^{π} equals 23.1406926 . . . , then i^{-2i} must also equal 23.1406926. . . . Who would have guessed that equality?

Many famous mathematicians and indeed those who write about mathematics often speak of the astonishing beauty and elegance that is found in mathematics.

For example, Martin Gardner once said, "All mathematicians share a sense of amazement over the infinite depth and mysterious beauty and usefulness of mathematics."[11]

Perhaps the great British mathematician Godfrey Harold Hardy (1877–1947) had equations such as these in mind when he said in his book *A Mathematician's Apology*: "The mathematician's patterns, like the painter's or poet's, must be beautiful. The ideas, like the colours or the words, must fit together in a harmonious way. Beauty is the first test: There is no permanent place in the world for ugly mathematics."[12]

Or perhaps the great British philosopher and mathematician Bertrand Russell (1872–1970) was contemplating these equations when he stated, "Mathematics rightly viewed contains not only truth but supreme beauty—a beauty cold and austere, like that of a sculpture."[13]

The great physicist Richard P. Feynman (1918–1988) once stated: "To those who do not know mathematics it is difficult to get across a real feeling as to the beauty, the deepest beauty, of nature . . . . If you want to learn about nature, to appreciate nature, it is necessary to understand the language that she speaks in."[14]

Feynman's remarks in turn remind me of Galileo's statement concerning the method in which nature operates: "The Book of Nature is written in mathematics."[15]

When I see beautiful equations, I think of what the great French mathematician and theoretical physicist Henri Poincaré (1854–1912), who said: "The mathematician does not study pure mathematics because it is useful; he studies it because he delights in it and he delights in it because it is beautiful."[16]

Mary Beth Ruskai (1944–), an American mathematician, once said, "We cannot hope that many children will learn mathematics unless we find a way to share our enjoyment and show them its beauty as well as its utility."[17]

When I had completed this chapter, I sent it to Dr. Cong, asking the good doctor for any comments he might wish to make. He sent the following details to me by e-mail.

Hi, Owen,

Hope you are keeping well.

I like your chapter on beautiful equations. Here are a few thoughts I have on the matter.

In the opinion of many mathematicians, number patterns are a reflection of the deep beauty and harmony that exists in nature. They also, of course, illustrate the beauty that is inherent in mathematics.

In many ways, a beautiful number pattern is like a beautiful sunset. It is something that is mysteriously beautiful. Why should we humans find a sunset—or indeed other natural phenomena—beautiful? Reality does not have to be that way. Reality could just as easily be such that we do not find these natural phenomena attractive, never mind beautiful. Why we humans find natural phenomena beautiful has baffled philosophers for centuries.

The great Hungarian mathematician Paul Erdős once asked: "Why are numbers beautiful? It's like asking why is Beethoven's Ninth Symphony beautiful. If you don't see why, someone can't tell you. I *know* numbers are beautiful. If they aren't beautiful, nothing is."[18]

One of the great theorems of mathematics is surely the Pythagorean theorem. Let the three sides of a right triangle equal a, b, and c (c being the longest side). The three lengths need not necessarily be integers. The theorem holds no matter what the lengths are, provided that they are real, positive numbers. The theorem simply states that if we build a square on the side of a, then build a square on the side of b, then the area of those two squares will equal in area a square built on the side of c. Mathematically this is expressed as $a^2 + b^2 = c^2$.

The simplest example in integers is $3^2 + 4^2 = 5^2$. It is easily proved that there are an infinite number of solutions in integers to this equation. The first six Pythagorean triples in which the hypotenuse is one more than the longer leg are as follows:

$$3^2 + 4^2 = 5^2$$
$$5^2 + 12^2 = 13^2$$
$$7^2 + 24^2 = 25^2$$
$$9^2 + 40^2 = 41^2$$
$$11^2 + 60^2 = 61^2$$
$$13^2 + 84^2 = 85^2$$
$$\ldots$$

Figure 16.27

The hypotenuses of these Pythagorean right triangles are 5, 13, 25, 41, 61, 85, 113, 145, 181, 221 . . . . Consider the following simple pattern:

$$(5 + 25)/13 = 2 \text{ plus a remainder of } 4$$
$$(13 + 41)/25 = 2 \text{ plus a remainder of } 4$$
$$(25 + 61)/41 = 2 \text{ plus a remainder of } 4$$
$$(41 + 85)/61 = 2 \text{ plus a remainder of } 4$$
$$(61 + 113)/85 = 2 \text{ plus a remainder of } 4$$
$$(85 + 145)/113 = 2 \text{ plus a remainder of } 4$$
$$\ldots$$

Figure 16.28

The following pattern is also beautiful:

$$(3 + 4) \cdot 5 = 6^2 - 1$$
$$(5 + 12) \cdot 13 = 15^2 - 2^2$$
$$(7 + 24) \cdot 25 = 28^2 - 3^2$$
$$(9 + 40) \cdot 41 = 45^2 - 4^2$$
$$(11 + 60) \cdot 61 = 66^2 - 5^2$$
$$(13 + 84) \cdot 85 = 91^2 - 6^2$$
$$(15 + 112) \cdot 113 = 120^2 - 7^2$$
$$\ldots$$

Figure 16.29

The areas of each of these right triangles are: 6, 30, 84, 180, 330, 546, 840, 1224 . . . .

Note the following pattern involving the triangular numbers:

$$6 = 2T_2$$
$$30 = 3T_4$$
$$84 = 4T_6$$
$$180 = 5T_8$$
$$330 = 6T_{10}$$
$$546 = 7T_{12}$$
$$840 = 8T_{14}$$
$$\ldots$$

Figure 16.30

The perimeters of these Pythagorean right are 12, 30, 56, 90, 132, and so on.

Consider the following pattern:

	Perimeter of Right Triangle
$4 + 8 = 12$	$(3, 4, 5,) = 12/1 = 12 = 3 \cdot 4$
$16 + 20 + 24 = 28 + 32 = 60$	$(5, 12, 13) = 60/2 = 30 = 5 \cdot 6$
$36 + 40 + 44 + 48 = 52 + 56 + 60 = 168$	$(7, 24, 25) = 168/3 = 56 = 7 \cdot 8$
$64 + 68 + 72 + 76 + 80 = 84 + 88 + 92 + 96 = 360$	$(9, 40, 41) = 360/4 = 90 = 9 \cdot 10$
. . .	

Figure 16.31

Note that if we add the perimeter of each triangle to the perimeter of the following right triangle, the following pattern, involving the triangular numbers and the odd numbers, emerges:

$$12 + 30 = 42 = T_9 - 3$$
$$30 + 56 = 86 = T_{13} - 5$$
$$56 + 90 = 146 = T_{17} - 7$$
$$\ldots$$

Figure 16.32

The following pattern involves the cubic numbers, the triangular numbers, and the natural numbers, 1, 2, 3, 4 . . . .

$$1^2 + 1^3 = 1 \cdot 2$$
$$2^2 + 2^3 = 3 \cdot 4$$
$$3^2 + 3^3 = 6 \cdot 6$$
$$4^2 + 4^3 = 10 \cdot 8$$
$$5^2 + 5^3 = 15 \cdot 10$$
$$6^2 + 6^3 = 21 \cdot 12$$
$$7^2 + 7^3 = 28 \cdot 14$$
$$\ldots$$

Figure 16.33

The next equation I give involves the square root of negative 1, $\sqrt{-1}$, which is usually denoted as i. The right-hand side of the equation includes the square root of the transcendental number known as e, $\sqrt{e}$. The number e equals 2.71828182 . . . . Its decimal expansion goes on forever. The right-hand side of the equation also includes the transcendental number known as π. It equals 3.14159265 . . . . Its decimal expansion also goes on forever. Yet the following equation shows how these three mathematical entities fit so neatly together.

$$i^{-i} = \sqrt{e}^{\pi}$$

Figure 6.34

The right-hand side of the equation can easily be calculated with a scientific calculator. Its value is 1.64872127 . . . raised to the power of π. This equals 4.8104773 . . . . Astonishingly, we find that i raised to the power of negative i is equal to a positive number!

In other words,

$$i^{-i} = 4.810477381 \ldots$$

Figure 16.35

On looking at such beautiful equations, you can only marvel at the wondrous, mysterious order that lies behind things!

Now I have some news for you. I am going away for a while, Owen. But I will return someday.

My little piece of advice to you, and to your readers, is keep that wondrous, mysterious order that is behind things in mind, and the fact that each of us human beings—a collection of atoms—can contemplate the cosmos, and the mathematical structure it apparently has. Atoms contemplating the existence of atoms! We are truly the universe's way of the universe being aware of its own existence. What an amazing fact!

Acknowledge the fact that it may well have been the case that none of us might have been here. We are here by mere chance. It is a remarkable coincidence that each of us is here, part of the universe. It is a great privilege to exist and to be able to witness and experience the great mystery of existence.

Appreciate life, my friend.

Finally, remember what James Barrie once said: "Those who bring sunshine to the lives of others cannot keep it from themselves."

I remain,

Your loyal friend,

Dr. Cong

NOTES

INTRODUCTION

1. Paul Menzer, *Crescendo of the Virtuoso: Spectacle, Skill, and Self-Promotion in Paris during the Age of Revolution*, Studies on the History of Society and Culture (Oakland: University of California Press, 1998).

CHAPTER 1: SOME WORDS ON THE *LO SHU* AND OTHER MAGIC SQUARES

1. Alex Bellos, "Magic Squares Are Given a Whole New Dimension," *Guardian* (Sunday newspaper in the United Kingdom), April 3, 2011.

2. Martin Gardner, "Magic Squares and Cubes," chap. 17 in *Time Travel and Other Mathematical Bewilderments* (New York: W. H. Freeman, 1988), pp. 213–25.

3. Many older sources in recreational mathematics state that in 1693 the French amateur mathematician Bernard Frénicle de Bessy enumerated that there were 880 order-four magic squares. For example, Maurice Kraitchik's *Mathematical Recreations* (first published in 1942 [New York: W. W. Norton]; republished in 1953 [New York: Dover, pp. 142–92]; and released in a second revised edition in 2006 [New York: Dover]) gives the number of order-four magic squares as 880. E. R. Berlekamp, J. H. Conway, and R. K. Guy (in *Winning Ways*, vol. 2 [London: Academic Press, 1982], pp. 778–83) give the number of different order-four magic squares as 880. In their book *The Power of Algorithms: Inspiration and Examples in Everyday Life* (Heidelberg, Germany: Springer-Berlin, 2013), Georgio Ausiello and Rosella Petreschi state that there are 880 order-four magic squares and 275,305,224 order-five magic squares.

The first analytical proof of this result was given by Kathleen Ollerenshaw and Herman Bondi in 1982. The number of order-five magic squares was first calculated in 1973 by Richard Schroeppel, using a computer program. The result was published for the first time in *Scientific American* in January 1976 (Eric W. Weisstein, "Magic Square," Wolfram MathWorld, mathworld.wolfram.com/MagicSquare.html [accessed March 1, 2015]).

That article also states that the number of magic squares of order six or above is unknown.

4. Martin Gardner, "The Magic of 3 × 3; the $100 Question: Can You Make a Magic Square of Squares?" *Quantum Magazine*, January/February 1996, pp. 24–26.

5. Lee Sallows, "The Lost Theorem," *Mathematical Intelligencer* 19, no. 4 (1997): 51–54.

6. Ibid.

7. Duncan Buell, *A Search for a Magic Hourglass*, July 1, 2004, http://www.multimagie.com/Buell.pdf.

8. A proof that a 3 × 3 magic square with distinct digits raised to any power greater than 2 is impossible is given by Henri Darmon and Loic Merel in a paper titled "Winding Quotients and Some Variants of Fermat's Last Theorem." This very technical paper was published (in English) on January 5, 2001, in *Journal für die reine und angewandte Mathematik* 490 (1997): 81–100.

9. The science magazine *Nature* carried a report of this on January 6, 2012.

10. *Wikipedia*, s.v. "Latin Square," https://en.wikipedia.org/wiki/Latin square.

11. The imaginary numerological friend I referred to in my chapter is Dr. Ming Cong. (His surname rhymes with the word *long*.)

CHAPTER 2: THE CALL OF THE PRIMES

1. *Wikipedia*, s.v. "Landau's Problems," https://en.wikipedia.org/wiki/Landau%27s_problems (last modified August 5, 2015).

2. Yitang Zhang Mac Arthur Foundation, http:// www.macfound.org (accessed August 5, 2015).

3. From 70 million to just 600, http://www.plusmaths.org (accessed August 4, 2015).

4. Ibid.

5. *Wikipedia*, s.v. "Vinogradov's Theorem," https://en.wikipedia.org/wiki/Vinogradov%27s_theorem (last modified August 6, 2015).

6. "Harald Helfgott Solves 'Odd' Goldbach Conjecture," May 15, 2013, http://www.cs.brandeis.edu/?p=465 (accessed August 1, 2015).

7. *Wikipedia*, s.v. "Chen's Theorem," https://en.wikipedia.org/wiki/Chen%27s_theorem (last modified August 13, 2015).

8. "Prime Numbers Hide Your Secrets," *Slate*, June 23, 2013, http://www.slate.com/articles/health_and_science/science/2013/06/online_credit_card_security_the_rsa_algorithm_prime_numbers_and_pierre_fermat.html (accessed September 1, 2015).

9. *Wikipedia*, s.v. "RSA Factoring Challenge," https://en.wikipedia.org/wiki/RSA_Factoring_Challenge (last modified September 5, 2015).

10. Ibid.

11. "Centre for the Popularisation of Mathematics," http://www.popmath.org.uk (accessed September 12, 2015).

12. http://www.history.mcs.st-and.ac.uk/Quotations/Einstein (accessed August 5, 2015).

13. David Wells, *The Penguin Dictionary of Curious and Interesting Numbers* (London: Penguin Books, 1986).

14. Marcus du Sautoy, *The Music of the Primes* (New York: HarperCollins, 2003), chap. 11.

15. Wolfram MathWorld, "Prime Number Theorem," http://mathworld.wolfram.com/topics/PrimeNumberTheorem.html (accessed January 5, 2016).

16. Paul Erdős, "Problems Worthy of Attack Prove Their Worth by Fighting Back," http://www.ucsb.edu (accessed September 2, 2015).

17. *Wikipedia*, s.v. "Mersenne Prime," https://en.wikipedia.org/wiki/Mersenne_prime (accessed September 2, 2015).

18. Ibid.

19. *Wikipedia*, s.v. "Fortunate Number," https://en.wikipedia.org/wiki/Fortunate_number (accessed September 2, 2015).

20. Joerg Arndt, On-line Encyclopedia of Integer Sequences, April 15, 2013. http://www.oeis.org (accessed September 1, 2015).

21. *Encyclopaedia Britannica*, s.v. "Jacques-Salomom Hadamard," http://www.brittanica.com (accessed October 10, 2015).

22. Big Primes, "Large List of Big Primes," http://www.bigprimes.net (accessed September 2, 2015).

23. *Wikipedia*, s.v. "Prime Number Theorem," https://en.wikipedia.org/wiki/Prime_number_theorem (accessed March 1, 2015).

24. George F. Simmons, *Calculus Gems: Brief Lives and Memorable Mathematics* (Washington, DC: Mathematical Association of America, 2007).

CHAPTER 3: SOME WORDS ON PYTHAGOREAN TRIPLES

1. "Natural Numbers: Exploring the Undesigned Intelligence of the Numberverse," www.naturalnumbers.org/bignum.html (accessed January 5, 2016).

2. "Questions and Answers: How Many Atoms Are in the Human Body?" Jefferson Lab, education.jlab.org/qa/mathatom_04.html (accessed January 5, 2016).

3. *Wikipedia*, s.v. "Augustine of Hippo," https://en.wikipedia.org/wiki/Augustine_of_Hippo (accessed August 15, 2015).

4. *Wikipedia*, s.v. "Russell's Teapot," https://en.wikipedia.org/wiki/Russell%27s_teapot (accessed August 3, 2015).

5. Richard Dawkins, *The God Delusion* (London: Bantam, 2006), p. 86.

6. *Encyclopaedia Britannica*, s.v. "Pythagorean Theorem," http://www.britannica.com/topic/Pythagorean-theorem (accessed March 1. 2015).

7. Martin Gardner, "The Pythagorean Theorem," chap. 16 in *Sixth Book of Mathematical Diversions from Scientific American* (Chicago: University of Chicago Press, 1971), p. 157.

8. Maurice Kraitchik, *Mathematical Recreations* (New York: Dover, 1953), pp. 101 and 102.

9. *Sue Liu, "Incircles Explained," NRICH, July 2001,* https://nrich.maths.org/1401 (accessed November 5, 2014).

10. Alexander Bogomolny, "Pythagorean Triples and Perfect Numbers," http://www.cut-the-knot.org/pythagoras/pythPerfect.shtml (accessed June 1, 2014).

11. Ibid.

12. *Wikipedia*, s.v. "Wiles Proof of Fermat's Last Theorem," https://en.wikipedia.org/wiki/Wiles'_proof_of_Fermat's_Last_Theorem (accessed March 2, 2015).

13. "Mass and Energy," http://www.profmattstrassler.com/articles-and.../mass-energy.../mass-and-energy/ (accessed June 4, 2015).

CHAPTER 5: THE FIBONACCI SEQUENCE

1. Parmanand Singh, "The So-Called Fibonacci Numbers in Ancient and Medieval India," *Historia Mathematica* 12, no. 3 (August 1985): 229–44.

2. Alfred S. Posamentier and Ingmar Lehmann, introduction to *The Fabulous Fibonacci Numbers* (Amherst, NY: Prometheus Books, 2007).

3. Martin Gardner, chap. 13 of *Mathematical Circus* (New York: Alfred Knopf, 1979).

4. Ibid.

5. *Wikipedia*, s.v. "Lucas Prime," https://en.wikipedia.org/wiki/Lucas_number (accessed September 29, 2015).

CHAPTER 7: THE IRRATIONAL NUMBER PHI

1. *Wikipedia*, s.v. "Gustav Fechner," https://en.wikipedia.org/wiki/Gustav_Fechner (accessed march 20, 2015).

2. *Wikipedia*, s.v. "Golden Spiral," https://en.wikipedia.org/wiki/Golden_spiral (accessed March 19, 2015).

3. Ibid.

4. Donald E. Simanek, "Fibonacci Flim-Flam," http://www.lhup.edu/~dsimanek/pseudo/fibonacc.htm (accessed January 5, 2016).

5. Pseudo Fibonacci, https://www.lhup.edu/~dsimanek/ (accessed March 20, 2015).

6. "*Li* or Principle: Stable Gravitational Orbits Possible Only in 3-Dimensional Space," www.swensonchangcasina.com/sources/stable_gravitational_orbits.php (accessed January 5, 2016).

7. Martin Gardner, "Gardner on Gardner: JPBM Communications Award Presentation," *Focus: The Newsletter of the Mathematical Association of America* 14, no. 6 (December 1994).

8. *Wikipedia*, s.v. "Positron," https://en.wikipedia.org/wiki/Positron (accessed November 18, 2104).

9. Math Quotes, "Heinrich Hertz," http://platonicrealms.com/quotes/Heinrich-Hertz (accessed October 25, 2104).

10. *Wikipedia*, s.v. "Paul Dirac," https://en.wikipedia.org/wiki/Paul_Dirac (accessed October 12, 2104).

11. Sean Carroll, "Einstein and Pi," http://www.preposterousuniverse.com/blog/2014/03/13/einstein-and-pi/ (accessed May 3, 2015).

CHAPTER 8: THE SQUARE ROOT OF –1

1. *Wikipedia*, s.v. "Gerolamo Cardano," https://en.wikipedia.org/wiki/Gerolamo_Cardano (accessed April 2, 2015).

2. *Wikipedia*, s.v. "Rafael Bombelli," https://en.wikipedia.org/wiki/Rafael_Bombelli (accessed April 2, 2015).

3. Tristan Needham, *Visual Complex Analysis* (Oxford, UK, and New York: Oxford University Press, 1999), pp. 16–19.

4. Freeman Dyson, "Birds and Frogs," *Notices of American Math. Society* 56 (2009): 212–23.

CHAPTER 10: THE SQUARE NUMBERS

1. Isaac Asimov, *Asimov's New Guide to Science* (London: Penguin Books, 1987), pp. 369.

2. *Wikipedia*, s.v. "Bobby Fischer," https://en.wikipedia.org/wiki/Bobby_Fischer (accessed June 1, 2014).

CHAPTER 11: THE TRIANGULAR NUMBERS

1. OddPerfectNumbers.org, "Various Papers on Odd Perfect Numbers," http://oddperfectnumber.org/papers.html (accessed April 28, 2014).

2. *Wikipedia*, s.v. "Srinivasa Ramanujan," https:www.//en.wikipedia.org/wiki/Srinivasa_Ramanujan (accessed January 10, 2015).

3. Dr. Math, "The Math Forum at Drexel," http://www.mathforum.org/library/ (accessed February 15, 2015).

4. Martin Gardner, *The Colossal Book of Short Puzzles and Problems* (New York: W. W. Norton, 2006), pp. 41, 50, and 51.

CHAPTER 12: THE TRANSCENDENTAL NUMBER KNOWN AS π

1. *Wikipedia*, s.v. "Uncertainty Principle," https://en.wikipedia.org/wiki/Uncertainty_principle (accessed June 5, 2015).

2. Sean Carroll, "Einstein and Pi," http://www.preposterousuniverse.com/blog/2014/03/13/einstein-and-pi/ (accessed January 14, 2015).

3. *Wikipedia*, s.v. "Stokes' Law," https://en.wikipedia.org/wiki/Stokes%27_law (accessed February 16, 2015).

4. Martin Gardner, *Martin Gardner's New Mathematical Diversions from Scientific American* (Chicago: University of Chicago Press, 1966), chap. 8, pp. 91–102.

5. *Wikipedia*, s.v. "Wallis Product," https://en.wikipedia.org/wiki/Wallis_product (accessed November 8. 2014).

6. Eric W. Weisstein, "Pi Formulas," Wolfram MathWorld, http://mathworld.wolfram.com/PiFormulas.html (accessed September 11, 2014).

7. Harvey Mudd College Math Department, "Math Fun Facts," http:// www.math.hmc.edu/funfacts (accessed January 5, 2016).

8. "Stirling," http://www.sosmath.com/ . . . us/sequence/ (accessed June 7, 2015).

9. "Pythagorean Right-Angled Triangles," sect. 9.6, "Pythagorean Triples and Pi," http:// www.maths.surrey.ac.uk/hosted-sites/R.Knott/Pythag/pythag.html (accessed January 5, 2016).

10. Boston University Physics Department, "Alternating Current," http://physics.bu.edu/~duffy/semester2/c20_AC.html (accessed January 5, 2016).

11. Siim Sepp, "Great Circles and Flight Paths," http://www.sandatlas.org/great-circles-and-flight-paths (accessed May 29, 2015).

12. Eric W. Weisstein, "Pi Approximations," Wolfram MathWorld, http://mathworld.wolfram.com/PiApproximations.html (accessed May 26, 2015).

CHAPTER 13: THE TRANSCENDENTAL NUMBER *e*

1. *Wikipedia*, s.v. "Jacob Bernoulli," https://en.wikipedia.org/wiki/Jacob_Bernoulli (accessed January 8, 2016).

2. *Wikipedia*, s.v. "Mathematical Constant," https://en.wikipedia.org/wiki/Mathematical_constant#Euler.27s_number_e (accessed July 5, 2015).

3. Edward Kasner and James Newman, *Mathematics and the Imagination* (New York: Dover, 2003).

4. Martin Gardner, "The Transcendental Number *e*," chap. 3 in *Further Mathematical Diversions* (Harmondsworth, Middlesex, England: Penguin Books, 1981).

5. Eli Maor, *e: The Story of a Number* (Princeton: Princeton University Press, 1994), p. 157.

6. H. Jerome Keisler, *Elementary Calculus: An Infinitesimal Approach* (New York: Dover, 2012).

7. *Encyclopaedia Britannica*, s.v. "Catenary Mathematics," http://www.search.eb.com/topic/catenary (accessed April 28, 2015).

8. Mohamed Amine Khamsi, "Newton's Law of Cooling," http://www.sosmath.com/diffeq/first/application/newton/newton.html (accessed June 26, 2014).

9. Maor, *e*, p. 103.

10. "Atmospheric Pressure and Altitudes Exponential Function" (table illustrating how the atmospheric pressure decays exponentially with altitude from what it is at the surface of the earth), http://www.regentsprep.org/regents/math (accessed January 5, 2016).

11. "Carbon 14 Dating," http://www.Carbon14Dating.mathcentral.uregina.ca/beyond/articles/ExpDecay/Carbon14 (accessed January 5, 2016). This page gives an example of how the carbon 14 method is applied.

12. "Modelling with Exponential and Logarithmic Functions," http://www.cims.nyu.edu/~kiryl/Precalculus/Problems/Section 4.6/Modeling (accessed January 5, 2016).

13. Ciara Curtin, "Fact or Fiction? Living People Outnumber the Dead," *Scientific American*, March 1, 2007.

14. Calvin C. Clawson, *Mathematical Mysteries: The Beauty and Magic of Numbers* (Cambridge, MA: Perseus Books, 1996), chap. 5, pp. 112–15.

15. *Wikipedia*, s.v. "Natural Logarithm," https://en.wikipedia.org/wiki/Natural_logarithm (accessed September 20, 2014).

16. "How Many Primes Are There?" http://www.primes.utm.edu/howmany (accessed January 5, 2016). Table 1 on this page gives the number of primes for successive powers of 10, up to 10^{25}.

17. Martin Gardner, "The Transcendental Number *e*," chap. 3 in *Further Mathematical Diversions* (Harmondsworth, Middlesex, England: Penguin Books, 1981).

18. "Permutations and Combinations," http://www.pas.rochester.edu/~stte/phy104-F00/notes-3.html (accessed November 2, 2014).

19. D. H. Lehmer, "The Asymptotic Evaluation of Certain Totient Sums," *American Journal of Mathematics* 22 (1900).

20. Calvin C. Clawson, *Mathematical Mysteries: The Beauty of Numbers* (Cambridge, MA: Perseus Books, 1996), pp 161–62.

21. Heinrich Dörrie, *100 Great Problems of Elementary Mathematics: Their History and Solution* (New York: Dover, 1965).

22. Wolfram MathWorld, "Continuous Distributions: Uniform Sum Distribution," http://mathworld.wolfram.com/topics/ContinuousDistributions.html (accessed January 5, 2016).

23. *Wikipedia*, s.v. "Mathematical Constant."

24. Ibid.

25. *Wikipedia*, s.v. "Stirling's Approximation," https://en.wikipedia.org/wiki/Stirling%27s_approximation (accessed July 3, 2014).

26. "Wolfram Alpha Examples: Sequences," http:// www.wolframalpha.com/examples/Sequences.htm (accessed January 5, 2016). Enter the expression and instantly obtain the limit.

27. Ibid. (accessed April 15, 2014).

28. Ibid. (accessed March 24, 2014).

29. *Wikipedia*, s.v. "Euler's Identity," https://en.wikipedia.org/wiki/Euler%27s_identity (accessed July 27, 2015).

30. Gardner, "Transcendental Number *e*."

31. "A Question Comparing π^e and e^π," http://www.math.stackexchange.com/questions/337565/a-question-comparing-pi (accessed January 5, 2016).

CHAPTER 14: PASCAL'S TRIANGLE

1. *Wikipedia*, s.v. "Pascal's Triangle," https://en.wikipedia.org/wiki/Pascal%27s_triangle (accessed August 28, 2015).

2. Alexander Bogomolny, "Pi in Pascal's Triangle," Cut the Knot, http://www.cut-the-knot.org/arithmetic/algebra/PiInPascal.shtml (accessed August 27, 2015).

3. Ibid.

4. William James, *Some Problems of Philosophy: A Beginning of an Introduction to Philosophy* (New York: Longman's, Green, 1911), p. 46; new edition published Lincoln, NE: University of Nebraska Press, 1996.

5. *Wikipedia*, s.v. "Perfect Number," https://en.wikipedia.org/wiki/Perfect_number (accessed August 29, 2015).

CHAPTER 15: SOME STRANGE AND REMARKABLE COINCIDENCES

1. *Wikipedia*, s.v. "American Airlines Flight 191," https://en.wikipedia.org/wiki/American_Airlines_Flight_191 (accessed July 5, 2014).

2. Sam Leith, "What Are You Doing Here? or Why Seemingly Amazing Coincidences Aren't So Unlikely after All," Mail Online, April 2, 2011, http://www.dailymail.co.uk (accessed January 5, 2016).

3. Associated Press, "Suspected Thief Caught in the Act—on Page One," *Oregonian*, December 14, 2007, http://blog.oregonlive.com/breakingnews/2007/12/suspected_thief_caught_in_the.html (accessed July 3, 2015).

4. John Allen Paulos, "The 9-11 Lottery Coincidence," October 6, 2015, http://www.abc.news.go.com/technology (accessed June 28, 2015).

5. Mark Singelais, "An Eerie Coincidence at Belmont Park on Sept. 11," http://blog.timesunion.com/horseracing/an-eerie-coincidence-at-belmont-park-on-sept-11/8305/ (accessed August 12, 2015).

6. "Odds Are Stunning Coincidences Can Be Expected," http://www.novinite.com/.../Odds (accessed January 5, 2016).

7. "White Blackbird Spotted in Dublin Garden," *RTE News*, January 26, 2012, http://www.rte.ie/news/special-reports/2012/0126/311525-blackbird/ (accessed April 12, 2015).

8. "Lost Daughter Right behind You, Dad," *Metro*, August 9, 2007, http://metro.co.uk/2007/08/09/lost-daughter-right-behind-you-dad-604317/ (accessed January 2, 2015).

9. "Wicklow Landowner Wins Right-of-Way Case," *RTE News*, February 8, 2012, http://www.rte.ie/news/2012/0208/312061-enniskerry/ (accessed March 29, 2015).

10. *Wikipedia*, s.v. "Littlewood's Law," https://en.wikipedia.org/wiki/Littlewood%27s_law (accessed March 4, 2015).

11. Kate Watson-Smythe, "Birthday to Remember for Family in a Million," *Independent*, December 22, 1997, http://www.independent.co.uk/news/birthday-to-remember-for-family-in-a-million-1290266.html (accessed February 22, 2015).

12. "Normal Numbers," Wolfram MathWorld, http://www.mathworld.wolfram.com (accessed January 5, 2016).

13. "Digits of Pi," http://www.angio.net/pi>/big (accessed January 5, 2016). This is an online resource to check if a specific sequence of digits is found within the first 200 million digits of pi.

14. "Brouwer and Heyting's Sequence," www.cecm.sfu.ca/~jborwein/brouwer.html (accessed November 22, 2015). The digits 9876543210 first occur beginning at decimal position 21,981,157,633.

15. Charles Q. Choi, "How the Moon Formed: Lunar Rocks Support Giant Impact Theory," June 5, 2014, http://www.space.com/26142-moon-formation-giant-impact-theory-support.html (accessed January 23, 2015).

16. Hanne Jakobsen, "What Would We Do without the Moon?" *ScienceNordic*, January 12, 2012, http://sciencenordic.com/what-would-we-do-without-moon (accessed January 23, 2015.

17. "Guardian Planets Jupiter and Saturn Shield the Earth from Catastrophic Comet Collisions," *Daily Mail*, July 31, 2009, www.dailymail.co.uk/sciencetech/article-1203405 (accessed November 21, 2015).

18. Dennis Overbye, "Jupiter: Our Cosmic Protector?" *New York Times*, July 25, 2009, http://www.nytimes.com/2009/07/26/weekinreview/26overbye.html?_r=0 (accessed January 5, 2015).

19. "Ancestor of All Placental Mammals Revealed," *Science*, March 29, 2015 http://news.sciencemag.org/plants-animals/2013/02/ancestor-all-placental-mammals-revealed (accessed January 5, 2016).

20. *Wikipedia*, s.v. "Multiverse," https://en.wikipedia.org/wiki/Multiverse (accessed May 22, 2015.

21. "Did the Moon Sink the *Titanic*?" *Daily Mail*, March 6, 2012, www.dailymail.co.uk/sciencetech/article-2110842 (accessed January 5, 2016).

22. Fergus J. Wood, *Tidal Dynamics: Coastal Flooding and Cycles of Gravitational Force* (Norwell, MA: Kluwer Academic Publishers, 1986).

23. *Wikipedia*, s.v. "Birthday Problem," https://en.wikipedia.org/wiki/Birthday_problem (accessed October 12, 2015).

24. Wolfram Alpha, "Online Birthday Problem Calculator," http:// www.wolframalpha.com/input/?i=birthday problem calculator&a=F (accessed January 5, 2016).

25. John Dyson, *The Prime Minister's Boat Is Missing* (Sydney, Australia: Angus and Robertson, 1974).

26. "4.5-Billion-Year-Old Meteorite Crashes into Paris Family Home," *Telegraph*, October 10, 2011, http://www.telegraph.co.uk/news/science/space/8818456/4.5-billion-year-old-meteorite-crashes-into-Paris-family-home.html (accessed January 5, 2016).

27. "Festive Bash for Christmas Twins," Herald Scotland, December 25, 1994, http://www.heraldscotland.com/news/12537726.Festive_bash_for_Christmas_twins/ (accessed August 28, 2015).

28. "Comette Family Home Damaged by Egg-Sized Meteorite," Guardian, October 10, 2011, http://www.theguardian.com/world/2011/oct/10/comette-family-home-damaged-meteorite (accessed January 4, 2016).

29. "The First Reactor," http://www.ost.gov/understanding (accessed August 30, 2015).

CHAPTER 16: BEAUTIFUL MATHEMATICAL EQUATIONS

1. *Wikipedia*, s.v. "Murray Gell-Mann" https://en.wikipedia.org/wiki/Murray_Gell-Mann (accessed October 15, 2015).

2. Murray Gell-Mann, "Beauty, Truth and . . . Physics?" TED Talk filmed March 2007, https://www.ted.com/talks/murray_gell_mann_on_beauty_and_truth_in_physics?language=en (accessed October 1, 2015).

3. *Wikipedia*, s.v. "Paul Erdős," https://en.wikipedia.org/wiki/Paul_Erd%C5%91s (accessed October 11, 2015.

4. Martin Gardner, *The Magic Numbers of Dr. Matrix* (Amherst, NY: Prometheus Books, 1985), p. 35. (The original equation by Everett W. Comstock, published by Gardner, had been contributed to *Recreational Mathematics Magazine* in April 1962.)

5. Philippe Deléham, http://www.oeis.org (A048473), February 23, 2014 (accessed November 16, 2015).

6. *Wikipedia*, s.v. "Pi," en.wikipedia.org/wiki/Pi (accessed January 5, 2016).

7. *Wikipedia*, s.v. "Euler's Identity," https://en.wikipedia.org/wiki/Euler%27s_identity (accessed November 21, 2015).

8. Ibid.

9. Jörg Arndt and Christoph Haenel, *π Unleashed* (Berlin, Germany: Springer, 2000), p. 224.

10. Ibid.

11. "Martin Gardner Quotes," www.azquotes.com/author/5341-Martin_Gardner (accessed January 5, 2016).

12. G. H. Hardy, *A Mathematician's Apology* (England: Cambridge University Press, 1940).

13. "Quotes Details, Bertrand Russell," http://www.quotationspage.com/quote/26185 (accessed January 5, 2016)..

14. Richard Feynman, *The Character of Physical Law* (Middlesex, England: Penguin Books, 1992); the first edition was published by the British Broadcasting Corporation (BBC) in 1965.

15. "The Book of Nature Is Written in the Language of Mathematics," The Renaissance Mathematics, https://thonyc.wordpress.com/.../the-book-of-nature-is-written-in-the- (accessed January 7, 2016).

16. "Is Mathematics Beautiful?" Cut the Knot, www.cut-the-knot.org/manifesto/beauty.shtml (accessed November 21, 2015).

17. William E. Tavernetti, "Mathematics and Beauty," www.math.ucdavis.edu/~etavernetti/Mathematics_and_Beauty.pdf (accessed November 21, 2015).

18. Paul Hoffman, *The Man Who Loved Only Numbers* (London: Fourth Estate, 1998), p. 44.

SELECT BIBLIOGRAPHY

Abbott, H. L., P. Erdős, and D. Hanson. "On the Number of Times an Integer Occurs as a Binomial Coefficient." *American Mathematical Monthly* 81 (1974): 256–61.

Anderson, Ken. *Coincidences: Chance or Fate*? London, UK: Blandford, 1995.

Arndt, Jörg, and Christopher Haenel. *π Unleashed*. Berlin, Germany: Springer, 1998.

Asimov, Isaac. "Exclamation Point." Chap. 3 in *Asimov on Numbers*. New York: Bell, 1982.

———. "A Piece of *Pi*." Chap. 6 in *Asimov on Numbers*. New York: Bell, 1982.

Balaguer, Mark. *Platonism and Anti-Platonism in Mathematics*. New York: Oxford University Press, 2001.

Beiler, Albert H. "Digits—and the Magic of 9." Chap. 8 in *Recreations in the Theory of Numbers: The Queen of Mathematics Entertains*. New York: Dover, 1964.

———. "On the Square." Chap. 15 in *Recreations in the Theory of Numbers: The Queen of Mathematics Entertains*. New York: Dover, 1964.

Crompton, Jenny. *Unbelievable: The Bizarre World of Coincidences*. London, UK: Michael O'Mara Books, 2013.

Derbyshire, John. *Prime Obsession: Bernhard Riemann and the Greatest Unsolved Problem in Mathematics*. New York: Plume, 2004.

Deza, Michel Marie, and Elena Deza. *Figurate Numbers*. 5 Toh Tuck Link, Singapore: World Scientific Publishing, 2012.

Fults, John Lee. *Magic Squares*. Chicago: Open Court, 1974.

Gardner, Martin. "Coincidence." Chap. 1 in *Knotted Doughnuts and Other Mathematical Entertainments*. New York: W. H. Freeman, 1986.

———. "Fibonacci and Lucas Numbers." Chap. 13 in *Mathematical Circus*. New York: Alfred A. Knopf, 1979.

———. *The Magic Numbers of Dr. Matrix*. Buffalo, NY: Prometheus Books, 1985. (See pages 23 and 24 to see how J. F. Kennedy's name is linked to the number 35; chapter 6 contains a nice curiosity concerning π.)

———. *More Mathematical Puzzles of Sam Loyd*. New York: Dover, 1960. (Puzzle number 146 in this wonderful book is "The Courier Problem"; this is a similar problem as that given in chapter 9 concerning the column of soldiers and the dog.)

———. "Pascal's Triangle." Chap. 15 in *Mathematical Carnival*. New York: Alfred A Knopf, 1975.

———. "Some New Discoveries about 3 × 3 Magic Squares." Chap. 23 in *A Gardner's Workout: Training the Mind and Entertaining the Spirit*. Natick, MA: A. K. Peters, 2001.

———. "Phi: The Golden Ratio." Chap. 8 in *More Mathematical Puzzles and Diversions*. Harmondsworth, Middlesex, England: Penguin Books, 1966. Pages 69–81.

———. "The Pythagorean Theorem." Chap. 16 in *Martin Gardner's Sixth Book of Mathematical Diversions from Scientific American*. Chicago: University of Chicago Press, 1971.

———. "The Square Root of 2." Chap. 2 in *A Gardner's Workout: Training the Mind and Entertaining the Spirit*. Natick, MA: A. K. Peters, 2001.

———. "Strong Laws of Small Primes." Chap. 12 in *The Last Recreations*. New York: Springer, 1997.

———. "3 × 3 Magic Squares." Chap. 22 in *A Gardner's Workout: Training the Mind and Entertaining the Spirit*. Natick, MA: A. K. Peters, 2001.

———. "The Transcendental Number *e*." Chap. 3 in *Further Mathematical Diversions*. Harmondsworth, Middlesex, England: Penguin Books, 1981.

———. "The Transcendental Number *Pi*." Chap. 8 in New *Mathematical Diversions from Scientific American*. Chicago: University of Chicago Press, 1966.

Green, Thomas M., and Charles L Hamberg. *Pascal's Triangle*. 2nd ed. Charleston, SC: Create Space, 2012.

Hand, David. *The Improbability Principle*. London, UK. Bantam, 2014.

Henle, Michael, and Brian Hopkins. "Sam Loyd's Courier Problem, with Diophantus, Pythagoras and Martin Gardner." Chap 27 in *Martin Gardner in the Twenty-First Century*. Washington, DC: Mathematical Association of America, 2011. This article originally appeared in *College Mathematics Journal* 39, no. 5 (November 2008): 387–91.

Hersh, Reuben. *What Is Mathematics, Really?* New York: Vintage Books, 1998.

Hoggatt, Verner E., Jr. *Fibonacci and Lucas Numbers*. Boston: Houghton Mifflin, 1969.

Holt, Jim. *Why Does the World Exist?* New York: W. W. Norton, 2012.

Kasner, Edward, and James Newman. *Mathematics and the Imagination*. New York: Dover, 2003.

Livio, Mario. *The Story of Phi: The World's Most Astonishing Number*. New York: Broadway Books, 2003.

Loomis, Elisha Scott. *The Pythagorean Proposition*. Reston, VA: National Council of Teachers of Mathematics, 1968.

Loyd, Sam. *Sam Loyd's Cyclopedia of 5,000 Puzzles Tricks and Conundrums with Answers*. New York: Franklin Bigelow Corporation, Morningside Press, 1914. Pages 315 and 382.

Maor, Eli. *e: The Story of a Number*. Princeton: Princeton University Press, 1994.

———. *The Pythagorean Theorem: A 4,000-Year History*. Princeton: Princeton University Press, 2007.

Moscovich, Ivan. *The Monty Hall Problem and Other Puzzles*. New York: Dover, 2011.

O'Shea, Owen. "Albrecht Dürer and His Magic Square." *Journal of Recreational Mathematics* 35, no. 4 (2006): 285–90.

———. "More Simple—but Little Known—Methods of Calculating Pythagorean Triples." *Journal of Recreational Mathematics* 37, no. 3 (2008): 187–91.

———. "New Curiosities on the Lo Shu." *Journal of Recreational Mathematics* 35, no. 1 (2006): 23–29.

———. "Simple—but Little Known—Methods of Calculating Pythagorean Triples. *Journal of Recreational Mathematics* 37, no. 1 (2008): 1–8.

Oglivy, C. Stanley, and John T. Anderson. "Continued Fractions." Chap. 10 in *Excursions in Number Theory*. New York: Dover, 1988.

Pasles, Paul C. *Benjamin Franklin's Numbers: An Unsung Odyssey*. Princeton, NJ: Princeton University Press, 2007.

Pickover, Clifford A. *A Passion for Mathematics: Numbers, Puzzles, Madness, Religion, and the Quest for Reality*. Hoboken, NJ: John Wiley and Sons, 2005. (Page 78 of this wonderful book contains a beautiful equation involving π.)

———. *Wonders of Numbers: Adventures in Mathematics, Mind and Meaning*. New York: Oxford University Press, 2003.

———. *The Zen of Magic Squares, Circles and Stars*. Princeton, NJ: Princeton University Press, 2002.

Plimmer, Martin, and Brian King. *Beyond Coincidence: Stories of Amazing Coincidences and the Mystery and Mathematics that Lie behind Them*. Cambridge, UK: Icon Books, 2004.

Posamentier, Alfred S. *Pi: A Biography of the World's Most Mysterious Number*. Amherst, NY: Prometheus Books, 2004.

———. *The Pythagorean Theorem: The Story of Its Power and Beauty*. Amherst, NY: Prometheus Books, 2010.

Posamentier, Alfred S., and Ingmar Lehmann. *The Glorious Golden Ratio*. Amherst, NY: Prometheus Books, 2011.

Ribenboim, Paulo. *The Little Book of Big Primes*. New York: Springer, 1991.

Rosenhouse, Jason. *The Monty Hall Problem: The Remarkable Story of Math's Most Contentious Brain Teaser*. New York. Oxford University Press, 2009.

Sautoy, Marcus du. *The Music of the Primes: Searching to Solve the Greatest Mysteries in Mathematics*. New York: HarperCollins, 2003.

Shapiro, Stewart. *Thinking about Mathematics*: *The Philosophy of Mathematics*. New York: Oxford University Press, 2000.

Sierpinski, Waclaw. *Pythagorean Triangles*. New York: Dover, 2003.

Singmaster, David. "How Often Does an Integer Occur as a Binomial Coefficient?" *American Mathematical Monthly* 78 (1971): 385–86.

Stewart, Ian. *Professor Stewart's Incredible Numbers*. London: Profile Books, 2015.

Thompson, Silvanus P., and Martin Gardner. *Calculus Made Easy*. Rev. ed. New York: St. Martin's, 1998. The original edition of *Calculus Made Easy* by Silvanus P. Thompson was published in 1910, 1914, and 1946, by MacMillan, London, UK.

Vaughan, Alan. *Incredible Coincidence: The Baffling World of Synchronicity*. New York: Ballantine Books, 1979.

Watkins, Matthew. *The Mystery of the Prime Numbers*: *Secrets of Creation*. Vol. 1. Illustrated by Matt Tweed. Dursley, Gloucestershire: Inamorata, 2010.

Weaver, Warren. *Lady Luck: The Theory of Probability*. New York: Doubleday, 1963.

Wells, David. *The Penguin Book of Numbers*. London: Penguin Books, 1987.

———. *Prime Numbers: The Most Mysterious Figures in Math*. Hoboken, NJ: John Wiley and Sons, 2005.

———. *The Penguin Dictionary of Curious and Interesting Numbers*. London: Penguin Books, 1986. (In "Root 2," pp. 34 and 35, David Wells describes a beautiful property of the $\sqrt{2}$ discovered by Roland Sprague.)

———. *The Penguin Dictionary of Curious and Interesting Numbers*. Rev. ed. London: Penguin Books, 1997.

INDEX